On Finiteness in Differential Equations and Diophantine Geometry

Volume 24

CRM MONOGRAPH SERIES

Centre de Recherches Mathématiques
Université de Montréal

On Finiteness in Differential Equations and Diophantine Geometry

Dana Schlomiuk
Andreĭ A. Bolibrukh
Sergei Yakovenko
Vadim Kaloshin
Alexandru Buium

Dana Schlomiuk
Editor

American Mathematical Society
Providence, Rhode Island USA

The work of D. Schlomiuk is partially supported by NSERC.

The work of A. A. Bolibrukh was partially supported by the grant "Leading Scientific Schools of the Russian Federation" (project no. 00-15-96107).

The research of S. Yakovenko was supported by the Israeli Science Foundation grant no. 18-00/1.

The work of V. Kaloshin is partially supported by the Sloan Dissertation Fellowship and the American Institute of Mathematics Five-Year Fellowship.

2000 *Mathematics Subject Classification.* Primary 34C07, 34C08, 34M35, 14G25; Secondary 11J91, 11J82, 12H05, 32G10, 34C10, 34M10, 34M50, 37C20, 37C27, 58K20.

For additional information and updates on this book, visit
www.ams.org/bookpages/crmm-24

Library of Congress Cataloging-in-Publication Data

On finiteness in differential equations and diophantine geometry / Dana Schlomiuk... [et al.], editor.

p. cm. — (CRM monograph series, ISSN 1065-8599 ; v. 24)

Includes bibliographical references.

ISBN 0-8218-2805-3 (alk. paper)

1. Differential-algebraic equations. 2. Diophantine analysis. I. Schlomiuk, Dana. II. Series.

QA372.5.O53 2005
515′.38—dc22 2005049858

∞ The paper used in this book is acid-free and falls within the guidelines
established to ensure permanence and durability.
This volume was submitted to the American Mathematical Society
in camera ready form by the Centre de Recherches Mathématiques.
Visit the AMS home page at http://www.ams.org/

10 9 8 7 6 5 4 3 2 1 10 09 08 07 06 05

We dedicate this volume to the memory
of the splendid mathematician and friend,
Andreĭ Andreevich Bolibrukh
who died on November 11, 2003.

Contents

Foreword

A workshop entitled "Asymptotic Series, Differential Algebra and Finiteness Problems in Nonlinear Dynamical Systems" was held at the Centre de recherches mathématiques (CRM) at the Université de Montréal from June 19 to July 7, 2000. This workshop was organized by Luc Bélair and myself. To enlarge our vision, a number of mathematicians working in model theory or working on finiteness problems in Diophantine geometry were invited to participate in this workshop. The diverse interests of the participants, the variety of methods used, as well as the format of the workshop (which allowed ample time for discussions) resulted in interesting scientific exchanges and also led to new results in this area of research.

The present monograph is based on four of the main series of lectures at this workshop and it also contains further research done after the workshop took place and which in some cases was stimulated by the scientific discussions which occurred at, or as a result of, the workshop.

An overview of the material in the monograph is given in the first chapter. It is meant to indicate how the four other chapters form a general picture of finiteness problems in differential equations and Diophantine geometry highlighting relations among these chapters.

This workshop was made possible by partial support by the following organizations: National Science and Engineering Council of Canada (NSERC), ministère de l'Éducation du Québec, Association des Universités partiellement ou entièrement de langue française (AUPELF). We would like to thank the director at that time of the CRM, Jacques Hurtubise, who gave his full support for the workshop and Louis Pelletier and Josée Laferrière for their splendid organization of the workshop. We would also like to thank Dr. Arthur Greenspoon who read the entire manuscript and made suggestions to improve the text. Many thanks are also due to Jean LeTourneux for overseeing the final work in the preparation of the manuscript and to Ina Lindemann for her advice. Special thanks are due to André Montpetit for the very fine work he has done in formatting it in this final form and of course to my colleague Luc Bélair for a wonderful cooperation in the organization of this workshop.

Dana Schlomiuk
October 2004

CHAPTER 1

Finiteness Problems in Differential Equations and Diophantine Geometry

D. Schlomiuk

In a finiteness problem we are asked to prove that a certain set of mathematical objects is finite or that it depends on a finite number of parameters. The term also admits generalizations. A finiteness problem was considered by Dulac in [**D**] (1923):

Prove that any real planar differential system

$$\frac{dx}{dt} = Q(x, y), \qquad \frac{dy}{dt} = P(x, y) \tag{1}$$

where P and Q are polynomials with real coefficients, has a finite number of limit cycles, i.e., of isolated periodic orbits.

In [**D**] Dulac claimed to have proved this but in the 1970's Dumortier began to suspect the presence of a gap in his proof. In 1981 Ilyashenko actually found an error [**I1**] and the search for a sound proof began. In the late 1980's Écalle and Ilyashenko, independently, gave proofs which they later published in two books [**É, I3**] which they wrote on this subject. As expected, asymptotic expansions intervene in both books but while Ilyashenko's proof has as a basic ingredient the analytic continuation in the complex domain of the Poincaré first return real maps associated with polycycles, Écalle's proof uses the Borel – Laplace transforms and his theory of resummation of divergent series. A polycycle (or in some texts a graphic) is a finite sequence of singular points p_i, $i = 1, \dots, l$, (with possible repetitions) and an associated finite sequence of connecting oriented orbits γ_i with α, respectively ω-limit points at p_i and p_{i+1} and γ_l connecting p_l with p_1. Both Écalle and Ilyashenko actually proved a more general theorem:

Every analytic vector field over a compact surface has a finite number of limit cycles.

In 1922 (a year before Dulac's work [**D**] was published), Mordell formulated another famous finiteness problem called the Mordell conjecture:

Prove that every algebraic curve of genus $g > 1$ over the rational numbers has a finite number of rational points.

Just like Dulac's problem, the Mordell conjecture was also without a proof for over sixty years, until Faltings solved this conjecture in [**F**] in 1983. Faltings' result is actually a more general theorem:

Let k be a number field and C an algebraic curve of genus g, $g > 1$, defined over k. Then C has a finite number of k-rational points.

Faltings received the Fields medal in 1986 at the ICM in Berkeley for his work.

It often happens that after a result is obtained, people will view it from different angles, simplify its proof and perhaps connect the result to other areas of mathematics.

A completely different proof of Faltings' theorem was given by Vojta in [**V**]. Vojta's proof, which uses Diophantine approximations, was later simplified by Bombieri in [**B1**]. An account of the Vojta – Bombieri proof, accessible to graduate students, can be found in Hindry – Silverman's book [**HS**].[1]

In his presentation [**M**] of the work of Faltings at the ICM 1986 in Berkeley, Mazur indicates that the analogue of Mordell's conjecture in the function field case was obtained by Manin (1963), Grauert (1965) with a different proof, Arakelov using an idea of Parshin (1971), and Zarhin (1974) also using the work of Parshin.

Mazur mentions the important role played in Faltings' proof of Mordell's conjecture by the analogy between number fields and function fields. It was by studying the proof of Parshin – Zarhin of the function field analogue of Mordell's conjecture that Faltings succeeded in discovering a method analogous to the one of Parshin – Zarhin which led to his proof of the Mordell conjecture in the number field context. Mazur mentions the important role played in the work of Faltings by the developments that occurred in algebraic geometry and number theory in the sixties and early seventies: "These developments which enter in an essential way in the work of Faltings encompass three grand themes, and Faltings proved the conjecture of Mordell by first establishing the truth of some other outstanding conjectures — fundamental to arithmetic algebraic geometry." The three grand themes mentioned by Mazur are: (1) The arithmetization of geometry and the geometrization of arithmetic; (2) Curves and Abelian varieties; and (3) Abelian varieties and Galois representations.

Finding different proofs not only opens the door for more people to understand a result, it even leads to new results as seen in Faltings' adaptation to the number theoretic setting of the Parshin – Zarhin method of proof in the function field analogue of the Mordell conjecture. Viewing a result from different angles and using powerful analogies between mathematical contexts are important elements of the creative process.

Although over ten years have passed since Écalle and Ilyashenko proved their theorem, no other proof, simpler or not, has appeared for their result. In [**S2**], Smale said that the proofs of Ilyashenko and Écalle "have still to be digested by the mathematical community." As this proof was the first major general result obtained on the second part of Hilbert's 16th problem, one naturally hopes that this situation will improve in the future.

The second part of Hilbert's 16th problem asks for the maximum $H(n)$ of the numbers of limit cycles and their positions occurring in planar vector fields defined by polynomials of degrees equal to n. We stress here that Hilbert's question is very precise, asking for the value of this maximum which is usually called the Hilbert number and is denoted by $H(n)$. The extreme difficulty of this problem is clear since after more than a century it remains open even for quadratic vector fields, i.e., for the case $n = 2$. Its difficulty stems from the fact that this problem has a *global* (even doubly global) character: for each n we need to consider the whole

[1]The author thanks A. Buium for providing this information.

class of systems defined by polynomials of degree n and for each such system we need to consider limit cycles in the whole plane.

In view of Écalle and Ilyashenko's Theorem (also called "the individual finiteness theorem" as it is about a given, fixed vector field) we now know that $H(n) \leq \aleph_0$. Is $H(n)$ finite?

The problem of proving that $H(n)$ is finite is called the finiteness part of Hilbert's 16th problem.[2] This is also a very hard problem. More than a century after Hilbert stated it, not even the finiteness part of his 16th problem in the quadratic case ($n = 2$) has been solved. In [**DRR**] Dumortier, Roussarie and Rousseau gave a program for proving the finiteness of $H(2)$. In [**DRR**] they reduced this problem (via the compactification of each one of the systems and of the parameter space (see [**R2, DRR**])) to the problem of proving the finite cyclicity[3] of all the polycycles and degenerate limit periodic sets (limit periodic sets were defined in [**FP, R2**]) involved in showing that $H(2)$ is finite and they find 121 such cases. Numerous articles were published on this problem and at this point the finite cyclicity has been proved for about 85 of the cases. However these articles do not deal with the most important cases which are the most degenerate ones. Since Hilbert's 16th problem, as well as its finiteness part, are global problems, the generic as well as the degenerate ones must be taken into account. The methods used for the proof of the 85 cases, which are basically methods of bifurcation theory, do not apply (as of now) in the case of the most degenerate ones of the limit periodic sets; for example, they do not apply for those surrounding a center and with a segment of the line at infinity filled up with singularities. In [**I4**] it is claimed that *all* the polycycles occurring in quadratic systems are listed in [**DRR**]. This is not true. The result given in [**DRR**] lists *only* the polycycles which are *involved in the proof of the finiteness of* $H(2)$; the global problem of listing all polycycles occurring in quadratic systems is still open.

As Hilbert's 16th problem has a global content, truly global methods for dealing with this problem must be developed. Naturally the first case where one needs to apply such global methods is the quadratic case. Here is what Hilbert had to say regarding hard problems:

In dealing with mathematical problems, specialization plays, as I believe, a still more important part than generalization. Perhaps in most cases where we seek in vain the answer to a question the cause of the failure lies in the fact that problems simpler and easier than the one in hand have been either not at all or incompletely solved.

For a discussion on the second part of Hilbert's 16th problem and global issues concerning this problem in the quadratic case see [**S1**]. Deep, important results on the global theory of quadratic systems are proved in [**Z1, Z2**]. Problems in the global theory of quadratic differential systems are also treated in [**NV, SV1, SV2, RS**].

For a survey where results for both the first and the second part of Hilbert's 16th problem are given and a long list of references is included, the reader interested in this problem may consult [**L2**]. In particular this list of references includes

[2]This problem was also called the *existential part* of Hilbert's 16th problem. A better name, suggested by Yakovenko in the chapter he wrote for this volume, is *existential finiteness* as it requires proving the existence of a finite bound for $H(n)$.

[3]This concept was defined in [**R2**]. See definitions of this notion in both Kaloshin's and Yakovenko's chapters in this volume.

significant contributions of Chinese mathematicians who have had a long tradition of work on this problem and which are absent from [**I4**].

While the second part of Hilbert's 16th problem still defies us, it gave birth to a number of finiteness problems, some of them called "weakened" or "restricted versions" of Hilbert's 16th problem, and they generated a large amount of research. Some of these have been solved, for example a problem formulated in the context of smooth, finite parameter generic families of vector fields with compact parameter space, on the 2-sphere. Apart from asymptotic expansions and methods of bifurcation theory, these results also involve geometrical methods and, in explicit or implicit ways, methods of differential algebra.

In the summer of the year 2000 a workshop entitled "Asymptotic Series, Differential Algebra and Finiteness Problems in Nonlinear Dynamical Systems" was held at the Centre de recherches mathématiques (CRM), at the Université de Montréal (June 19 – July 7, 2000). The workshop, which was organized by Luc Bélair and Dana Schlomiuk, had as its goals to facilitate exchanges of ideas among the participants about all these finiteness problems, to expose the participants to a variety of viewpoints and techniques in such problems and to stimulate new research in this area.

Five series of lectures were delivered, four of them of five lectures each and a series of four lectures on "Logarithmic-exponential series and o-minimality." These were given respectively by Andrey Bolibrukh, Alexandru Buium, Vadim Kaloshin, Sergey Yakovenko and Lou van den Dries. There were also eleven one hour lectures of communications by other participants. The workshop opened with a series of lectures entitled "Asymptotic series and differential equations" given by Andrey Bolibrukh, famous for his work on Hilbert's 21st problem. Bolibrukh lectured on linear ordinary differential equations with meromorphic coefficients, a subject with a much older history than the nonlinear theory and one which has applications in the nonlinear case. In particular such applications were encountered in Yakovenko's lectures where a number of finiteness theorems on nonlinear systems were discussed.

It was thought appropriate to widen the program by also including a series of lectures on finiteness problems in another area of mathematics. The speaker for this series was Alexandru Buium, whose lectures were entitled "Differential Algebraic Geometry and Diophantine Geometry." Roughly speaking Diophantine geometry is the part of mathematics which studies sets of solutions of systems of polynomial equations over number fields or function fields. An example of a finiteness result in diophantine geometry is Faltings' theorem mentioned earlier. Buium obtained proofs of finiteness theorems in Diophantine geometry using methods of differential algebra. The use of such methods formed a link among the five series of lectures as they were encountered in explicit or implicit ways in all of them.

Another area of mathematics in which contributions to diophantine geometry were made (mostly in the function field case) is model theory, a branch of mathematical logic. In model theory in the 1980's the notion of o-minimal structure (= order-minimal)[4] was defined by axioms which imply finiteness properties encountered in semialgebraic and subanalytic geometry. It was thought appropriate to include in the workshop a series of lectures on o-minimal structures. This lecture series entitled "Logarithmic-exponential series and o-minimality" was

[4]This notion was introduced by Pillay and Steinhorn in [**PS**] based on an idea of van den Dries [**vdD**].

delivered by Lou van den Dries who suggested the possibility that o-minimality may eventually provide some explanation of why finiteness properties hold. In his lectures he also gave a formal construction, analogous to the construction of the "trigèbre de transséries" [**É**] of Écalle. Many participants at the workshop were exposed for the first time to these model-theoretic methods and their existing connections to the main theme of the meeting. The content of these lectures is covered by publications of van den Dries written on this topic. Some of his publications are listed in the website page of Lou van den Dries, whose address is: `http://www.math.uiuc.edu/People/vddries.html`

Apart from this introduction, this monograph is made up of four chapters written by four of the main speakers of the workshop around our main theme of finiteness problems. These chapters are not the exact content of their lectures. They were written later and some contain further research, done after the workshop took place, which in some instances was stimulated by the scientific exchanges which occurred at, or as a result of, the workshop.

Bolibrukh's chapter in this book, entitled "Linear Differential Equations, Fuchsian Inequalities and Multiplicities of Zeros," begins with a very brief introduction to the analytic theory of linear systems of differential equations with meromorphic coefficients, part of his lectures at this workshop. Bolibrukh then presents his new results concerning generalized Fuchsian inequalities for such systems, which he used in order to obtain estimates on multiplicities of zeroes for the components of solutions of such systems. The results concerning Fuchsian inequalities were obtained after the workshop took place and as Bolibrukh says towards the end of his opening section, were "inspired by very fruitful discussions during the school with D. Schlomiuk, S. Yakovenko, N. Novikov, A. Buium and with other participants." They could be used for the estimation of the number of zeroes of Abelian integrals, intervening in the *linearized (or tangential) Hilbert's 16th problem*, which is discussed in the chapter written by Yakovenko in this book.

The chapter of this monograph written by Yakovenko is entitled "Quantitative Theory of Ordinary Differential Equations and Tangential Hilbert's 16th Problem." The goal of this chapter, which is written in a lively and rather informal style, is to discuss a number of finiteness problems for the number of zeros of analytic functions which are defined by ordinary differential equations. The main focus is on the *tangential Hilbert's 16th problem* on zeros of complete Abelian integrals.

As Yakovenko affirms, the connection between limit cycles of polynomial vector fields and isolated zeros of Abelian integrals is of a limited nature. However this connection "justifies formulation of several finiteness problems."

In finiteness problems there are two aspects: the qualitative and the quantitative. The qualitative part involves proofs of existence of finite bounds in specific finiteness problems. Such problems are called by Yakovenko *existential finiteness problems*. The quantitative part is the search for specific estimates of bounds and for as good an estimate as is possible. Yakovenko calls *constructive finiteness problem* a problem asking to find an explicit bound in case where the existence of a finite bound is already proved.

A synopsis of "weakened" or "restricted" versions of Hilbert's 16th problem appears in Table 1 entitled "Various flavors of the Hilbert's 16th problem" where Yakovenko lists the *individual finiteness*, the *existential finiteness*, the *constructive finite cyclicity*, the *constructive finiteness* and corresponding results whenever we have them. This chapter is focused on the *tangential Hilbert's 16th problem* which as Yakovenko says in Section 1.3.2 is the problem of "computing and majorizing the number of zeros of Abelian integrals." The existential finiteness in this case was proved and it is called the Varchenko – Khovanskiĭ theorem (Theorem 3 of Section 1.4.2 of the chapter).

The connection between Abelian integrals and linear differential equations is explained by Yakovenko in the second section of his chapter. This connection forms a link between the nonlinear theory of differential equations and the linear theory encountered in the preceding chapter written by Bolibrukh. In Table 2, a partial synopsis is given for the *constructive tangential Hilbert's 16th problem.* More precisely, Abelian integrals form a finitely generated module over the ring of polynomial functions of t, and generators of this module satisfy a system of first order linear ordinary differential equations with rational coefficients in t, called the Picard – Fuchs equations. The Picard – Fuchs linear equations are also present in the third section where quasialgebraicity of function fields is defined and motivation for this notion is given. The quantitative theory of differential equations is discussed in the fourth section of the chapter where explicit quantitative results are given. In the last section connections are made between this work on the quantitative theory of differential equations and the Riemann – Hilbert problem. This is the 21st problem on Hilbert's list which was solved in the negative by a counter-example given by Bolibrukh in 1990 [**B2**].

Another finiteness problem which sprang from Hilbert's 16th problem is the *Hilbert – Arnol'd problem* which asks for a proof that in a generic finite parameter family of smooth vector fields on the sphere S^2 with compact parameter space B, the number of limit cycles is uniformly bounded by a natural number. This problem is at the core of Kaloshin's chapter of this book and, as he explains, this problem can be reduced to the study of the bifurcations of limit periodic sets. The existential finiteness for the cyclicity of elementary polycycles (i.e., polycycles with with elementary singularities — meaning by this that they have at least one nonzero eigenvalue) was proved by Ilyashenko and Yakovenko [**IY**]. They proved the finiteness of the cyclicity $E(k)$ of a nontrivial elementary polycycle occurring in a generic k-parameter smooth family of vector fields on the sphere. In his chapter of this book, Kaloshin explains his beautiful proof of his result (the main theorem in this chapter) giving the first known explicit and sufficiently general estimate for this cyclicity $E(k)$. The proof is based on a variation of the important geometric reduction method of Khovanskiĭ. Kaloshin extends Khovanskiĭ's method with interesting additional arguments which allow him to obtain the explicit bound. Kaloshin's chapter goes beyond finiteness problems and results on planar vector fields, including results for estimates of bifurcation numbers attached to polycycles in n-dimensional phase space. These theorems are of a quantitative nature, complementing the quantitative results obtained in the previous two chapters by Bolibrukh and Yakovenko.

The closing chapter of this book, entitled "Finiteness Results in Differential Algebraic Geometry and Diophantine Geometry," written by Alexandru Buium, is about the interaction between differential algebraic geometry, and Diophantine geometry over function fields. As Buium indicates in [**B4**] differential algebraic geometry was founded by Ritt and Kolchin. Its relation to algebraic geometry is analogous to the relation existing between differential algebra and commutative algebra. In differential algebraic geometry the role of algebraic equations is replaced by algebraic differential equations. The material in this chapter is based on Buium's work on this subject. Rather than writing a faithful transcript of his lectures in the workshop, Buium gives here less background than in his lectures but covers more ground, including topics not previously discussed in the lectures such as the *abc* conjecture and more.

In a series of papers Buium developed a differential algebraic method which he used to prove diophantine results over function fields. In particular Buium proved a quantitative version of the Manin – Mumford conjecture in [**B5**] using differential algebraic techniques. The following is the Manin – Mumford finiteness conjecture:

Let X be a smooth projective curve over $\mathbb{C}$ of genus $g > 1$ and consider its embedding into its Jacobian $\mathrm{Jac}(X)$. *Then the number of torsion points of X is finite.*

This conjecture was first proved by Raynaud in [**R1**] by methods of algebraic geometry. For a full history of this conjecture and of its various proofs see [**HS, B3**]. Using model theoretic methods, Hrushovski proved a generalized version of the Manin – Mumford conjecture ([**H1**]).

A conjecture which contains both the Manin – Mumford and Mordell's conjecture is the conjecture of Lang [**L1**] (also called the Mordell – Lang conjecture). Perhaps its simplest form, which appears in [**B4**], is the following:

Let Y be a compact Riemann surface of genus $g \geq 2$ embedded in a complex compact torus A and let Γ be a subgroup of finite rank of A. Then the intersection $Y \cap \Gamma$ is finite.

As Buium mentions in his chapter, Lang's conjecture (stated more generally than above) was proved by Faltings and Hindry. The first goal of Buium's chapter in this book is to prove a special case of the conjecture of Lang, namely the Geometric Lang conjecture. This he does by using a "bigger subgroup that becomes a geometric object in a richer geometry, the so-called *differential algebraic geometry*." This idea was the starting point for the model theoretic proof of Hrushovski of the Lang conjecture over function fields in any characteristic [**H2, B3**]). Buium also explains how the differential algebraic method can be pushed to work "beyond" the geometric Lang conjecture, becoming relevant to the geometric case of the *abc* conjecture.

Finiteness problems and results along with differential algebraic methods (used in explicit or implicit ways) are connecting elements linking together the chapters in this book.

At this point it seems appropriate to close with the following words taken from Hilbert's 1900 address [**H**]:

Mathematical science is in my opinion an indivisible whole, an organism whose vitality is conditioned upon the connection of its parts. For with all the variety of mathematical knowledge, we are still clearly conscious of the similarity of the logical devices, the relationship *of the* ideas *in mathematics as a whole and the numerous analogies in its different departments. We also notice that, the farther a mathematical theory is developed, the more harmoniously and uniformly does its construction proceed, and unsuspected relations are disclosed between hitherto separate branches of the science. So it happens that, with the extension of mathematics, its organic character is not lost but only manifests itself the more clearly.*

References

[B1] E. Bombieri, *The Mordell conjecture revisited*, Ann. Scuola Norm. Sup. Pisa, Cl. Sci. (4) **17** (1990), no. 4, 615–640.

[B2] A. A. Bolibrukh, *The Riemann–Hilbert problem*, Uspekhi Mat. Nauk **45** (1990), no. 2(272), 3–47 (Russian); English transl., Russian Math. Surveys **45** (1990), no. 2, 1–58.

[B3] É. Bouscaren, *Théorie des modèles et conjecture de Manin–Mumford (d'après Ehud Hrushovski)*, Séminaire Bourbaki, Vol. 1999–2000, Astérisque, vol. 276, Soc. Math. France, Paris, 2002, pp. 137–159.

[B4] A. Buium, *Géométrie algébrique différentielle et géométrie diophantienne*, La Gazette des sciences mathématiques du Québec **16** (1994), no. 2, 1–8.

[B5] ———, *Geometry of p-jets*, Duke Math. J. **82** (1996), no. 2, 349–367.

[vdD] L. van den Dries, *Remarks on Tarski's problem concerning* $(\mathbb{R}, +, \cdot, \exp)$, Logic Colloquium '82 (Florence, 1982) (G. Lolli, G. Longo, and A. Marcja, eds.), Stud. Logic Found. Math., vol. 112, North-Holland, Amsterdam, 1984, pp. 97–121.

[D] H. Dulac, *Sur les cycles limites*, Bull. Soc. Math. France **51** (1923), 45–188.

[DRR] F. Dumortier, R. Roussarie, and C. Rousseau, *Hilbert's 16th problem for quadratic vector fields*, J. Differential Equations **110** (1994), no. 1, 86–133.

[É] J. Écalle, *Introduction aux fonctions analysables et preuve constructive de la conjecture de Dulac*, Actualités Math., Hermann, Paris, 1992.

[F] G. Faltings, *Endlichkeitssätze für abelsche Varietäten über Zahlkörpern*, Invent. Math. **73** (1983), no. 3, 349–366.

[FP] J.-P. Françoise and C. C. Pugh, *Keeping track of limit cycles*, J. Differential Equations **65** (1986), no. 2, 139–157.

[H] D. Hilbert, *Mathematical problems.*, Bull. Amer. Math. Soc. **8** (1902), 437–479; reprinted, Bull. Amer. Math. Soc. (N.S.) **37** (2000), no. 4, 407–436.

[HS] M. Hindry and J. H. Silverman, *Diophantine geometry: An introduction*, Grad. Texts in Math., vol. 201, Springer-Verlag, New York, 2000.

[H1] E. Hrushovski, *The Manin–Mumford conjecture and the model theory of difference fields*, Ann. Pure Appl. Logic **112** (2001), no. 1, 43–115.

[H2] ———, *The Mordell–Lang conjecture over function fields*, J. Amer. Math. Soc. **9** (1996), 667–690.

[I1] Yu. S. Ilyashenko, *Singular points and limit cycles of differential equations on the real and complex plane*, Sci. Res. Comp. Centre, Pushchino, 1982.

[I2] ———, *Finiteness theorems for limit cycles*, Uspekhi Mat. Nauk **45** (1990), no. 2(272), 143–200 (Russian); English transl., Russian Math. Surveys **45** (1990), no. 2, 129–203.

[I3] ———, *Finiteness theorems for limit cycles*, Transl. Math. Monogr., vol. 94, Amer. Math. Soc., Providence, RI, 1991.

[I4] ———, *Centennial history of Hilbert's 16th problem*, Bull. Amer. Math. Soc. (N.S.) **39** (2002), no. 3, 301–354.

[IY] Yu. Ilyashenko and S. Yakovenko, *Finite cyclicity of elementary polycycles in generic families*, Concerning the Hilbert 16th Problem, Amer. Math. Soc. Transl. Ser. 2, vol. 165, Amer. Math. Soc., Providence, RI, 1995, pp. 21–95.

[L1] S. Lang, *Division points on curves*, Ann. Mat. Pura Appl. (4) **70** (1965), 229–234.

[L2] J. Li, *Hilbert's 16th problem and bifurcations of planar vector fields*, Internat. J. Bifur. Chaos Appl. Sci. Engrg. **13** (2003), no. 1, 47–106.

[M] B. Mazur, *On some of the mathematical contributions of Gerd Faltings*, Proceedings of the International Congress of Mathematicians. Vol. 1 (Berkeley, CA, 1986), Amer. Math. Soc., Providence, RI, 1987, pp. 7–15.

[NV] I. Nikolaev and N. Vulpe, *Topological classification of quadratic systems at infinity*, J. London Math. Soc. (2) **55** (1997), no. 3, 473–488.

[PS] A. Pillay and C. Steinhorn, *Definable sets in ordered structures*. I, Trans. Amer. Math. Soc. **295** (1986), no. 2, 565–592.

[R1] M. Raynaud, *Courbes sur une variété abélienne et points de torsion*, Invent. Math. **71** (1983), no. 1, 207–233.

[R2] R. Roussarie, *A note on finite cyclicity property and Hilbert's 16th problem*, Dynamical Systems (Valparaiso, 1986), Lecture Notes in Math., vol. 1331, Springer, Berlin, 1988, pp. 161 – 168.

[RS] R. Roussarie and D. Schlomiuk, *On the geometric structure of the class of planar quadratic differential systems*, Qual Theory Dyn. Syst. **3** (2002), no. 1, 93 – 121.

[S1] D. Schlomiuk, *Aspects of planar polynomial vector fields: global versus local, real versus complex, analytic versus algebraic and geometric*, Normal Forms, Bifurcations and Finiteness Problems in Differential Equations (Montréal, 2002) (Yu. Ilyashenko and C. Rousseau, eds.), NATO Sci. Ser. II Math. Phys. Chem., Kluwer Acad. Publ., Dordrecht, 2004, pp. 471 – 509.

[SV1] D. Schlomiuk and N. Vulpe, *Geometry of quadratic differential systems in the neighborhood of infinity*, Technical Report CRM-2701, Centre de recherches mathématiques, Université de Montréal, 2001.

[SV2] ———, *Geometry of quadratic differential systems in the neighborhood of infinity*, J. Differential Equations, to appear.

[S2] S. Smale, *Mathematical problems for the next century*, Mathematical Intelligencer **20** (1998), no. 2, 7 – 15.

[V] P. Vojta, *Siegel's theorem in the compact case*, Ann. of Math. (2) **133** (1991), no. 3, 509 – 548.

[Z1] P. G. Zhang, *On the distribution and number of limit cycles for quadratic systems with two foci*, Acta Math. Sinica **44** (2001), 37 – 44 (Chinese).

[Z2] ———, *On the distribution and number of limit cycles for quadratic systems with two foci*, Qual. Theory Dyn. Syst. **3** (2002), no. 2, 437 – 463.

CHAPTER 2

Linear Differential Equations, Fuchsian Inequalities and Multiplicities of Zeros

A. A. Bolibrukh

Abstract. A brief introduction to the analytic theory of linear systems of ordinary differential equations with meromorphic coefficients is presented as well as the recent results on generalized Fuchsian inequalities for such systems. Estimates for the multiplicity of zeros for components of solutions to linear systems obtained with the help of Fuchsian inequalities are given.

1. Introduction

This article is devoted to linear systems of ordinary differential equations with meromorphic coefficients. Linear systems appear in various domains of mathematics: in mathematical physics (the theory of isomonodromic deformations), in problems of integrability of Hamiltonian systems, in the linearized Hilbert's 16th problem, in analytic number theory, etc.

We give here a brief introduction to some aspects of the theory. Our main goal is the following: to present recent results concerning generalized Fuchsian inequalities for systems of linear differential equations and to deduce estimates for multiplicities of zeros for components of solutions to such systems with the help of these inequalities.

These results can be used for the estimations of the numbers of zeros of Abelian integrals appearing in the linearized Hilbert's 16th problem (see the articles of S. Yakovenko [**Y**], V. Kaloshin [**K**] on this subject, presented in this volume and [**NY**]) and for questions of algebraic dependence in analytic number theory as well (see [**BB, N**]).

Fuchsian inequalities turn out to be very useful for inverse problems of the analytic theory of differential equations, such as the problem of Birkhoff's standard form and in a number of applications of the theory.

Fuchsian inequalities give effective estimates for the sum of exponents of a system (of a scalar equation) over all singular points. Recall that the exponents $\{\beta_i^j\}$ of an equation at a singular point a_i are just the asymptotics of solutions to the equation at the point in the case of a regular singularity. At an irregular singular point the exponents coincide with polynomial asymptotics of solutions after elimination of the exponential part. The Fuchsian inequality in the case of a scalar Fuchsian equation is well known. It is the classical Fuchs relation (see [**I, H, CL**])

$$\sum_{i=1}^{n}\sum_{j=1}^{p}\beta_i^j = \frac{(n-2)p(p-1)}{2},$$

where n is the number of singular points, and p is the order of the equation.

The Fuchsian relation (39) for a scalar equation with irregular singularities was obtained in 1985 by D. Bertrand and G. Laumon in [**B**]. For systems of differential equations with regular singular points it was known that the sum of exponents is nonpositive [**B1**], but only recently have effective estimates been established by E. Corel. First he obtained the Fuchsian inequality (19) for the regular singular case in [**C1**] and then he did the same for the general case of irregular singular points in [**C2**] (estimate (29) in this article). To prove these results E. Corel uses algebraic methods. In this article we present elementary analytic proofs in Propositions 3, 4 and in Theorem 5.

The structure of the article is the following. In Section 2 we give a detailed description of systems with regular singular points with the emphasize on the Levelt valuation method. The presentation follows [**L, B1**] with some important simplifications concerning the using of the so-called Smith decomposition, which provides the possibility to transform a system to an upper triangular one.

Section 3 is devoted to irregular singular points. We deduce the form of a formal solution in the generic nonresonant case and we formulate the corresponding result in the general case. We don't discuss the very important questions concerning the k-summability of formal solutions. Here we refer to classical works [**MR, R, B1**], to a beautiful textbook [**B2**] and to a very well-written work [**LR**]. The central statement of the section is Proposition 4, which is used later to prove the generalized Fuchsian inequality.

In Section 4 we present all known Fuchsian inequalities. And Section 5 is devoted to estimates of multiplicities of zeros for components of solutions to a system of linear differential equations. Here we follow the schema of [**B3**].

The first part of the article represents partially the lectures that the author gave during the school on asymptotic series in Montreal. Another part (concerning multiplicities of zeros and Fuchsian inequalities) was inspired by very fruitful discussions during the school with D. Schlomiuk, S. Yakovenko, D. Novikov, A. Buium and with other participants.

The author is very grateful to the organizers of the school for the excellent organization of the conference which provided the possibility for such discussions.

2. Systems with regular singular points

Consider a system

$$\frac{dy}{dz} = B(z)y \tag{1}$$

of p linear differential equations on the Riemann sphere with singular points a_1, $\dots, a_n$. Denote by r_i the Poincaré rank of the system at the point a_i. This means that in a neighborhood of this point the matrix of the system is represented as

$$B(z) = \frac{B^i_{-r_i-1}}{(z-a_i)^{r_i+1}} + \cdots + \frac{B^i_{-1}}{z-a_i} + B^i_0 + \cdots, \quad B^i_{-r_i-1} \neq 0. \tag{2}$$

Recall that *a singular point a_i is called regular singular for the system* (1) *if all solutions have at most polynomial growth as z tends to zero remaining in some sector with the vertex at a_i*. (Since solutions to the system are multivalued functions one needs such a definition of polynomial growth.) In the opposite case the singular point is called *irregular*.

System (1) is called *Fuchsian at* a_i if the coefficient matrix of the system has a simple pole at a_i.

Consider a nonsingular point z_0 and a fundamental matrix $Y(z)$ of the space of solutions to system (1) (i.e., a matrix whose columns constitute a basis in the space of solutions to the system) in a neighborhood of the point. This matrix admits an analytic continuation along a loop γ in $\overline{\mathbb{C}} \setminus \{a_1, \dots, a_n\}$ with its origin at z_0. And we get a matrix $\widetilde{Y}(z)$, which must be connected with the initial one as follows:

$$Y(z) = \widetilde{Y}(z) G_\gamma,$$

where $G_\gamma \in \mathrm{GL}(p; \mathbb{C})$ (because the columns of these two matrices form two bases in the space of solutions to the system). This correspondence generates the *monodromy representation*

$$\chi\colon \pi_1(\overline{\mathbb{C}} \setminus \{a_1, \dots, a_n\}, z_0) \to \mathrm{GL}(p; \mathbb{C}) \tag{3}$$

of the system. Its image is called *the monodromy group* of the system and it is generated by *local monodromy matrices* $G_1, \dots, G_n$ corresponding to loops going around each singular point. These matrices satisfy a priori just one relation

$$G_1 \cdots G_n = I.$$

EXAMPLE 1. Consider a system of the form (1)

$$\frac{dy}{dz} = \begin{pmatrix} 1/z & 1 \\ 0 & 0 \end{pmatrix} y$$

with the Fuchsian singular point $z = 0$. A fundamental matrix $Y(z)$ of the space of solutions to the system has the form

$$Y(z) = \begin{pmatrix} z & z \ln z \\ 0 & 1 \end{pmatrix}.$$

It is clear that the point $z = 0$ is regular singular, because the rate of growth of the function $\ln z$ (more precisely of its arbitrary branch) is smaller than the rate of $1/z$ as z tends to 0 remaining in some sector with the vertex at 0.

For the system of one equation

$$\frac{dy}{dz} = -\frac{y}{z^2}$$

the point $z = 0$ is irregular singular, because its general solution is of the form $y = ce^{1/z}$.

It is well known that *every Fuchsian singular point is regular singular.*

Indeed, represent the matrix $B(z)$ of system (1) in the form $B(z) = B_0(z)/z$, where $B_0(z)$ is holomorphic in the unit disk D with the center at zero. From the inequality $\big|\|a\| - \|b\|\big| \le \|a - b\|$ it follows that

$$\big|d\|y(z)\|\big| \le \|dy(z)\|.$$

Let $\max_D \|B_0(z)\| = K$. From system (1) one has for $z \in \dot{D}$

$$\big|d\|y(z)\|\big| \le K\|y(z)\|\frac{|dz|}{|z|}, \quad \left|\frac{d\|y\|}{\|y\|}\right| \le K\frac{|dz|}{|z|}. \tag{4}$$

For every point $z_0 = \rho_0 e^{i\varphi} \in \dot{D}$ consider the segment J beginning at z_0 and with the end at $z_1 = e^{i\varphi}$. After integration of inequality (4) along J one gets

$$\left|\int_J \frac{d\|y\|}{\|y\|}\right| \leq \int_J \left|\frac{d\|y\|}{\|y\|}\right| \leq \int_J K\frac{|dz|}{|z|},$$

$$\big|\ln\|y(z_1)\| - \ln\|y(z_0)\|\big| \leq K\int_{\rho_0}^{1}\frac{d\rho}{\rho} = -K\ln\rho_0.$$

Thus,

$$\|y(z_0)\| \leq \|y(z_1)\|\,|z_0|^{-K}.$$

Since the set of values of every branch of the solution $y(z)$ is bounded on the boundary of the disk D, the latter inequality completes the proof.

The following example shows that the inverse statement is not true (there exist systems with regular singular points which are not Fuchsian).

EXAMPLE 2. The system

$$\frac{dy}{dz} = \begin{pmatrix} 0 & 1 \\ 1/z^2 & -1/z \end{pmatrix} y$$

is not Fuchsian at $z = 0$, nevertheless this point is regular singular for the system, because its fundamental matrix has the form

$$Y(z) = \begin{pmatrix} z & 1/z \\ 1 & -1/z^2 \end{pmatrix}.$$

Let now the singular point a_1 coincide with zero. Consider a monodromy matrix G_1 of system (1) at zero and denote it by G. Denote by E the matrix $(\frac{1}{2}\pi i)\ln G$. Let us choose the eigenvalues ρ^j of the matrix E such that the following inequalities hold

(5) $$0 \leq \operatorname{Re}\rho^j < 1.$$

Define the matrix z^E as follows:

$$z^E = \exp(E\ln z) = I + E\ln z + \cdots + E^n \ln^n z/n! + \dots .$$

Then under the analytic continuation along the loop γ_1 going around zero the matrix z^E goes to

$$\exp\big(E(\ln z + 2\pi i)\big) = z^E\exp(2\pi i E) = z^E G,$$

which means that the matrix z^E has the same monodromy as the initial matrix $Y(z)$. Thus, the following statement holds:

LEMMA 1. *The fundamental matrix $Y(z)$ has the following decomposition in a neighborhood of the point $z = 0$:*

(6) $$Y(z) = M(z)z^E,$$

where $M(z)$ is a single-valued matrix function.

Moreover, the point zero is regular singular if and only if the matrix $M(z)$ is meromorphic at zero.

The second statement of the lemma follows from the fact that if $\varepsilon > 0$, then the function $z^{E+\varepsilon I}$ tends to zero as z tends to zero remaining in any sector with the vertex at zero. Indeed, due to this fact the polynomial growth of Y is equivalent to the polynomial growth of M. Since M is uniform the latter condition means that the point zero is a pole for this function.

EXAMPLE 3. The fundamental matrix $Y(z)$ of the system from Example 1 has the following decomposition:

$$Y(z) = \begin{pmatrix} z & 0 \\ 0 & 1 \end{pmatrix} z^{\left(\begin{smallmatrix} 0 & 1 \\ 0 & 0 \end{smallmatrix}\right)}.$$

Let the function (multivalued) $y(z)$ have a polynomial growth at zero.

DEFINITION 1. The valuation of the function y at zero is the following integer:

$$\varphi(y) := \sup\left\{ k \in \mathbb{Z} \;\middle|\; \forall \lambda < k \; \frac{y(z)}{|z|^\lambda} \to 0, \text{ when } z \in S,\ z \to 0 \right\}$$

for every sector S with the vertex at zero. For $y = 0$ we define $\varphi(0) := \infty$.

EXAMPLE 4. The valuation of a meromorphic function at zero equals to the order of this function at zero:

$$\varphi\left(\frac{1}{z} \ln z\right) = -1; \quad \varphi(\sqrt{z}) = 0; \quad \varphi\begin{pmatrix} 1 + z^2 \ln z \\ 1/z + \ln z \end{pmatrix} = -1.$$

There is another way to introduce a valuation on the space of solutions to a regular system (1). It follows from Lemma 1 that every solution $y(z)$ can be presented in the form of a *finite logarithmic sum*

$$y(z) = \sum_{j,l \in \sigma} f_{jl}(z) z^{\rho_j} (\ln z)^{b_l}, \tag{7}$$

where f_{jl} are Laurent series with finite principal parts, the numbers ρ_j satisfy relation (5), the numbers b_l are nonnegative integers and all pairs (ρ_j, b_l) are distinct.

It is a simple exercise to prove that

the valuation of a solution $y(z)$ at zero is equal to the minimum over orders of the series f_{jl} at $z = 0$.

Consider the space of solutions to system (1) in a neighborhood of the singular point $z = 0$. Denote by σ^* the monodromy operator at zero (this operator maps every solution into its analytic continuation along a loop going around zero in the counterclockwise direction).

PROPOSITION 1. *The valuation φ determines a map $\varphi\colon X \to \mathbb{Z} \cup \infty$ with the following properties*:

(a) $\varphi(y_1 + y_2) \geq \min(\varphi(y_1), \varphi(y_2))$, *and if* $\varphi(y_1) \neq \varphi(y_2)$, *then one obtains equality*;
(b) $\varphi(cy) = \varphi(y)$ *for every* $c \in \mathbb{C} \setminus \{0\}$;
(c) $\varphi(\sigma^* y) = \varphi(y)$.

PROOF. The properties (a), (b) follow immediately from the definitions. Let us prove property (c).

Since $\sigma^* z^a = \exp(2\pi i a) z^a$ and $\sigma^* \ln z = \ln z + 2\pi i$, then from the form of the logarithmic sum it follows that

$$\varphi(\sigma^* y) \geq \varphi(y).$$

(The action of σ^* cannot lead to the appearance of functions $f_{jl}(z)$ with smaller orders at zero).

But the same is true for the operator $(\sigma^*)^{-1}$. Thus in the chain

$$\varphi(y) = \varphi\big((\sigma^*)^{-1} \circ \sigma^* y\big) \geq \varphi(\sigma^* y) \geq \varphi(y)$$

all inequalities become equalities. □

It follows from the properties (a) and (b) that φ takes on just a finite number of values $\infty > \psi^1 > \cdots > \psi^m$ on X and it determines the filtration

$$0 \subset X^1 \subset \cdots \subset X^m = X \tag{8}$$

of the space X by vector subspaces

$$X^k = \{y \in X \mid \varphi(y) \geq \psi^k\}, \qquad k = 1, \dots, m.$$

Due to the property (c) the operator σ^* preserves this filtration.

Consider a basis $e_1^1, \dots, e_{k_1}^1$ of the space X^1 such that the operator σ^* restricted to the space has an upper triangular matrix in this basis, complete it by a basis $e_1^1, \dots, e_{k_1}^1, e_1^2, \dots, e_{k_2}^2$ of the space X^2 with the same property, etc.

The constructed basis $(e) = (e_1, \dots, e_p)$ of the space X has the following properties:

(1) the valuation φ takes all its values with all multiplicities on the elements of the basis;
(2) $\varphi(e_{l+1}) \leq \varphi(e_l)$, $l = 1, \dots, p-1$;
(3) the matrix G of the operator σ^* in this basis has upper triangular form.

Such a basis is called a *Levelt basis.*

Consider the space X of solutions to system (1) with a regular singular point at zero and consider a Levelt basis (e) of X. Denote by G the matrix of the operator σ^* in this basis and denote by $A = \operatorname{diag}(\varphi^1, \dots, \varphi^p)$ the matrix of valuations of the basis (e), i.e., $\varphi^l = \varphi(e_l)$. Consider the matrix E from (5).

The following statement follows immediately from the definitions.

LEMMA 2. *The matrices $z^A G z^{-A}$ and $z^A E z^{-A}$ are holomorphic at the point $z = 0$. The valuation of the function $z^A z^E z^{-A}$ at zero equals zero.*

The following statement is a central one in the theory of systems with regular singular points. It was presented first in [**L**].

THEOREM 1. *For the fundamental matrix $Y_e(z)$ constructed from a Levelt basis (e) of a system with a regular singular point at zero the following decomposition holds:*

$$Y_e(z) = U(z) z^A z^E, \tag{9}$$

where the matrix $U(z)$ is holomorphic at $z = 0$.

PROOF. It follows from Lemma 1 that the function $U(z)$ is meromorphic. Consider the number $r = \max_i \operatorname{Re} \rho^i$, $0 < \varepsilon < (1-r)/2$. To prove the theorem it is sufficient to prove that

$$\lim_{z \to 0} U(z) z^{r+2\varepsilon} = 0.$$

(The limit is considered here for z from a sector with the vertex at zero as above.)

One has

$$U(z) z^{r+2\varepsilon} = Y_e(z) z^{-E} z^{-A} z^{r+2\varepsilon} = N_1(z) N_2(z),$$

where

$$N_1(z) = Y_e(z) z^{-A+\varepsilon I}, \quad N_2(z) = z^A z^{-E} z^{-A} z^{r+\varepsilon}.$$

Since the jth column of the matrix $N_1(z)$ equals

$$e_j(z) z^{-\varphi(e_j)+\varepsilon},$$

from the definition of the valuation it follows that $\lim_{z\to 0} N_1(z) = 0$.

From the choice of the number r one gets that the diagonal elements of the matrix $-E+rI$ have nonnegative real parts. Thus, as in the second part of Lemma 2 we obtain

$$\varphi\left(z^{A} z^{-E+rI} z^{-A}\right) = 0.$$

Therefore,

$$\lim_{z\to 0} N_2(z) = \lim_{z\to 0} z^{A} z^{-E+rI} z^{-A} z^{\varepsilon} = 0. \qquad \square$$

Using similar arguments one can prove the following statement

PROPOSITION 2. *Let a fundamental matrix $Y(z)$ of system (1) with the regular singular point at zero have the form*

$$Y(z) = U(z) z^{\Lambda} z^{E},$$

where $U(z)$ is holomorphic at zero, Λ is a diagonal integer valued matrix, and E is an upper triangular matrix with the property (5).

If the matrix $z^{\Lambda} E z^{-\Lambda}$ is holomorphic at zero, then the trace of the matrix Λ is less than or equal to the sum of the valuations of the system at zero.

PROOF. Let us prove that for every $\varepsilon > 0$ one has that the matrix $Y(z) z^{-\Lambda} z^{\varepsilon}$ tends to zero as z tends to zero remaining in a sector with the vertex at zero. Indeed,

$$Y(z) z^{-\Lambda} z^{\varepsilon} = U(z) z^{\Lambda} z^{E} z^{-\Lambda} z^{\varepsilon},$$

but $U(z)$ is holomorphic at zero and due to the conditions of the theorem and Lemma 2 the remainder of the expression tends to zero.

Thus, from the definition of the valuation it follows that the entry λ_{ii} of the matrix Λ is less than or equal to the valuation of the ith column of the matrix Y. And the same is true for the trace of Λ and for the sum of the valuations. $\square$

EXAMPLE 5. For the system from Example 2 the theorem gives the following decomposition:

$$Y(\tilde{z}) = \begin{pmatrix} z & z \\ 1 & -1 \end{pmatrix} z^{\left(\begin{smallmatrix} 0 & 0 \\ 0 & -2 \end{smallmatrix}\right)},$$

because $E = 0$ in this case.

And for the system from Example 1 one has

$$Y(\tilde{z}) = \begin{pmatrix} 1 & 0 \\ 0 & 1 \end{pmatrix} z^{\left(\begin{smallmatrix} 1 & 0 \\ 0 & 0 \end{smallmatrix}\right)} z^{\left(\begin{smallmatrix} 0 & 1 \\ 0 & 0 \end{smallmatrix}\right)}.$$

Note here that the matrix $U(z)$ from the last example is holomorphically invertible at zero and this is not the case for the previous system. It turns out that exactly this property distinguishes Fuchsian systems among all systems with regular singular points. More precisely the following theorem (which is due to Levelt) holds.

THEOREM 2. *System (1) with a regular singular point $z = 0$ is Fuchsian if and only if in decomposition (9)*

$$Y_e(z) = U(z) z^{A} z^{E}, \tag{10}$$

for the fundamental matrix $Y_e(z)$ constructed from a Levelt basis of the space of solutions X, the matrix $U(z)$ is holomorphically invertible at zero.

For what follows we need the following technical lemma (which is often called the Smith decomposition).

LEMMA 3. *For every matrix $M(z)$ holomorphically invertible in some neighborhood of the point zero and meromorphic at zero there always exist a matrix $P(z)$ and a matrix $V(z)$ such that*

(i) *$P(z)$ is upper triangular and rational in z;*
(ii) *$V(z)$ is holomorphically invertible in a neighborhood of zero;*
(iii) *the following decomposition holds:*

$$M(z) = V(z)P(z). \tag{11}$$

If the matrix $M(z)$ is holomorphic at zero, then $P(z)$ is a polynomial in z.

PROOF. Let the order of the pole of the matrix function $M(z)$ be equal to m. Replace the matrix M by the matrix $\widetilde{M}(z) = M(z)z^m$ which is holomorphic at zero. Denote by N_1 the matrix $\widetilde{M}(0)$. If this matrix is nondegenerate, then one has nothing to prove. If $\det N_1 = 0$, then multiplying the matrix N_1 on the right-hand side by an appropriate upper triangular matrix S_1 with units on the main diagonal one can always obtain a matrix N_1S_1 with ith zero column for some i. Thus,

$$\widetilde{M}(z)S_1 = M_1(z)z^{K_1},$$

where the matrix M_1 is holomorphic at zero and all entries of the matrix K_1 are equal to zero except the entry with the indices i, i which is equal to one.

If the matrix $N_2 = M_1(0)$ is degenerate, then applying the same procedure to the matrix N_2 we get matrices S_2, K_2, $M_2(z)$, etc. And in a finite number of steps we arrive at the following decomposition:

$$\widetilde{M}(z) = M_n(z)z^{K_n}S_n^{-1}z^{K_{n-1}}S_{n-1}^{-1}\cdots z^{K_1}S_1^{-1} = V(z)P(z),$$

where n is the order of the function $\det \widetilde{M}(z)$ at zero, $V(z) = M_n(z)$ is holomorphically invertible at zero and $P(z)$ is an upper triangular polynomial in z.

For the initial matrix $M(z)$ one gets

$$M(z) = V(z)Q(z), \quad Q(z) = P(z)z^{-m}.$$ □

REMARK 1. Let the matrix $M(z)$ be a formal meromorphic series at zero whose formal determinant does not vanish outside of zero. Then the previous lemma holds with the following changes. One must replace the holomorphically invertible matrix $V(z)$ by an invertible formal Taylor series in z.

Now we have all the ingredients to prove Theorem 2.

PROOF. Let us prove the necessity.

Due to Lemma 3 and relation (10), the Levelt fundamental matrix $Y_e(z)$ of system (1) has the following decomposition at zero:

$$Y_e(z) = V(z)P(z)z^A z^E, \tag{12}$$

where the matrix $V(z)$ is holomorphically invertible at zero and $P(z)$ is an upper triangular matrix polynomial in z.

Under the transformation $Y' = V(z)^{-1}Y_e$ the initial system goes to a system

$$\frac{dy'}{dz} = \frac{B_0'(z)}{z}y' \tag{13}$$

with the upper triangular matrix $B_0'(z)$ holomorphic at zero and with the fundamental matrix of the form

$$Y'(z) = P(z)z^A z^E. \tag{14}$$

Since this transformation is holomorphically invertible, it is evident that the latter system has the same exponents at zero as the initial one. To prove the theorem it is sufficient to prove that the diagonal elements of the matrix $P(z)$ don't vanish at zero. We shall prove this by induction with the respect to the size of the system.

For $p = 1$ the theorem is evidently true. Suppose it is proved for all systems of size smaller than p. We prove it for a Fuchsian system of size p (and of form (13)).

Since the subsystem of our system obtained by intersections of the first $p-1$ rows and columns of (13) is Fuchsian and has size $p-1$, from the induction hypothesis one has that the first $p-1$ elements of the main diagonal of the matrix $P(z)$ don't vanish at zero. Let the element with indices p, p of the matrix vanish at zero. Using this assumption we shall arrive at a contradiction and thus we shall prove the theorem for systems of size p.

From (13) one has

$$B_0'(z) = z\frac{dP}{dz}P^{-1} + P(A + z^A E z^{-A})P^{-1},$$

$$B_0'(z)P(z) = z\frac{dP}{dz} + P(z)(A + z^A E z^{-A}), \tag{15}$$

$$B_0'(0)P(0) = P(0)L(0), \quad L(z) = A + z^A E z^{-A}. \tag{16}$$

Note here that diagonal elements (eigenvalues) of the matrix $L(0)$ coincide with the diagonal elements of $A + E$ and are equal to $\beta^1, \dots, \beta^p$, where $\beta^i = \varphi^i + \rho^i$ and $\varphi^1 \geq \dots \geq \varphi^p$ while the real parts of ρ^i satisfy inequalities (5). The numbers β^i are called *the exponents of the system at zero.*

Due to the assumption the matrix $P(0)$ has rank $p-1$ and its kernel is generated by a vector $c = (c_1, \dots, c_p)^{\mathrm{t}}$ with $c_p = 1$ (the sign "t" means transposition). It follows from (16) that c is the eigenvector for the matrix $L(0)$ with the eigenvalue β^p. Let $\beta^{p-l} = \dots = \beta^p$ and $\beta^{p-l-1} \neq \beta^p$. Then among the components of c only the last $l+1$ components are not equal to zero. In this case we also get that $\varphi^{p-l} = \dots = \varphi^p$.

Consider the upper triangular matrix $S = I + C$, where all entries of the matrix C equal zero except the last column which is equal to $(0, \dots, 0, -c_{p-l}, \dots, -c_{p-1}, 0)^{\mathrm{t}}$. Then the matrix S commutes with A and for the fundamental matrix $\widetilde{Y} = Y'S^{-1}$ one gets

$$\widetilde{Y} = P'(z)z^A z^{E'}, \quad P'(z) = P(z)S^{-1}, \quad E' = SES^{-1}.$$

Since $c' = Sc = (0, \dots, 0, 1)$ and the vector c' is in the kernel of the matrix $P'(0)$ one gets that the last column of $P'(0)$ is equal to zero and thus

$$P'(z) = P''(z)z^K, \quad K = \operatorname{diag}(0, \dots, 0, 1)$$

with a holomorphic matrix P''.

Since due to (16) (rewritten for P' and $L'(z) = A + z^A E' z^{-A}$) the vector c' is the eigenvector for $L'(0)$, one gets that the last column of $L'(0)$ coincides with $\beta_p c'$. Therefore, the matrix $z^{A+K}E'z^{-A-K}$ is still holomorphic at zero. So, the Levelt decomposition for system (13) can be rewritten as follows:

$$\widetilde{Y}(z) = P''(z)z^{A+K}z^{E'}$$

and thus from Proposition 2 it follows that

$$\operatorname{tr}(A+K) \leq \operatorname{tr} A$$

and we arrive at a contradiction.

Now let us prove the sufficiency part of the theorem.

If the matrix U from (10) is holomorphically invertible at zero, then here as well as in (16), we get that the matrix $zB(z)$ has the form

$$zB(z) = z\frac{dU}{dz}U^{-1} + U\big(A + z^A E z^{-A}\big)U^{-1} \tag{17}$$

and thus it is holomorphic at zero. Therefore, the system is Fuchsian. □

Note here that from (17), for the matrix $B_0(z) = zB(z)$ one gets

$$B_0(0) = U(0)L(0)U(0)^{-1}$$

and therefore

the exponents of the Fuchsian system (1) *coincide with the eigenvalues of the residue of the coefficient matrix* $B(z)$ *at zero.*

Instead of Levelt bases one can also consider other bases, which we'll call weakly Levelt bases. They can be defined as follows.

If the monodromy operator σ^* has a single eigenvalue, then the notion of a weakly Levelt basis coincides with that of the usual Levelt basis. If this is not the case, then decompose the space X of solutions into a direct sum of root subspaces $X = X_1 \oplus \cdots \oplus X_s$, corresponding to distinct eigenvalues $\lambda^1, \dots, \lambda^s$ of the monodromy operator. Denote by σ_i^* the restriction of σ^* to X_i. The valuation φ restricted to X_i generates a filtration of this subspace similar to (8) and the operator σ_i^* preserves the filtration. For every X_i consider a corresponding Levelt basis. The basis formed by the union of the obtained bases over all X_i is called a *weakly Levelt basis* of the space X.

It is a simple exercise to prove that a *weakly Levelt basis of* X *is associated with the filtration* (8) *of the whole space* X. In other words the valuation φ takes all its values on a weakly Levelt basis (with corresponding multiplicities).

The matrix G of the monodromy operator σ^* has a block diagonal form $G = \operatorname{diag}(G_1, \dots, G_s)$ in a weakly Levelt basis and thus Lemma 2 holds in this case as well as Theorems 1 and 2, whose proofs repeat the presented proof.

Fuchsian system (1) is called *nonresonant* at the singular point if all differences between the eigenvalues of the residue of its coefficient matrix are either zero or noninteger numbers.

It follows immediately from the construction of a weakly Levelt basis that if the system is nonresonant at zero, then for every i all valuations of the root subspace X_i coincide. Thus, it follows from the block diagonal form of the matrix E in (12) that matrices A and E commute and thus in (16) one has $L(z) = L(0) = A + E$. So,

$$\exp\big(2\pi i L(0)\big) = \exp(2\pi i E) = G$$

and therefore, it follows from the relation noted below (17) that

if system (1) *is nonresonant at zero, then the monodromy matrix of the system at zero is conjugate to the matrix* $\exp\big(2\pi i B_0(0)\big)$.

From the fact that in the nonresonant case the matrices A and E commute, it also follows that

in the nonresonant case every Fuchsian at zero system (1) *is analytically equivalent at zero to the Euler system*

$$\frac{dy}{dz} = \frac{E + A}{z} y.$$

(This equivalence is given by the gauge transformation $y = U(z)y'$.) In the resonant case there is a special procedure (presented in [**CL**]) which also provides the possibility to calculate the local monodromy from the coefficient matrix.

The notions of monodromy, valuation and the corresponding filtrations can also be introduced for a scalar linear differential equation

$$u^{(p)} + q_1(z)u^{(p-1)} + \cdots + q_p(z)u = 0 \tag{18}$$

with regular singular points as well. Lemma 2 and Theorem 1 also holds for such an equation (upon replacing the matrix U by the corresponding vector row).

Equation (18) is called *Fuchsian at zero* if for every i the function $v_i(z) = z^i q_i(z)$ is holomorphic at zero. It turns out that instead of systems for scalar equations the notions of a regular singular point and of a Fuchsian point coincide (see [**CL, B1**]).

How do we calculate exponents of a Fuchsian scalar equation at a singular point? For this purpose it is necessary to consider the Euler equation associated with the Fuchsian one (18)

$$z^p u^{(p)} + v_1(0)z^{p-1}u^{(p-1)} + \cdots + v_i(0)z^{p-i}u^{(p-i)} + \cdots + v_p(0)u = 0.$$

(The latter equation is obtained by replacing the coefficients $v_i(z)$ of the Fuchsian equation by their values at the singular point. On the other hand one can treat a Fuchsian equation as a holomorphic deformation of an Euler equation, where the constant coefficients of the latter equation are replaced by holomorphic functions.)

It turns out (see again [**I, CL, B1**]) that

the exponents of a Fuchsian scalar equation (18) *are the roots of its* indicial equation

$$z^p \lambda^p + v_1(0)z^{p-1}\lambda^{p-1} + \cdots + v_i(0)z^{p-i}\lambda^{p-i} + \cdots + v_p(0)\lambda = 0.$$

Let us return to systems with regular singular points. What is the connection between the value of the Poincaré rank and the order of the function $\det U(z)$ from (9) at zero? The following statement, which is due to E. Corel [**C1**] takes place.

PROPOSITION 3. *For the order at zero of the function* $\det U(z)$ *from* (9), *the following inequalities hold*:

$$r \leq \operatorname{ord}_0 \det U(z) \leq \frac{p(p-1)}{2} r, \tag{19}$$

where r is the Poincaré rank of the system at zero.

PROOF. Consider again decomposition (12) for the Levelt fundamental matrix of the system and transform the system to the system

$$\frac{dy'}{dz} = B'(z)y' \tag{20}$$

with the fundamental matrix $Y' = V(z)^{-1}Y_e$. Now Y' has the form (14) and we must estimate the order of the determinant of P at zero.

System (20) has upper triangular form. Moreover, in the series (2) of B' at zero, one has that the matrices $B'_{-r-1}, \dots, B'_{-2}$ are nilpotent. This follows from the fact that the matrix P is upper triangular polynomial, and thus the diagonal entries of $(dP/dz)P^{-1}$ have simple poles at zero.

Therefore, the shearing transformation

$$\widetilde{Y} = z^C Y', \quad \text{where } C = \operatorname{diag}\big((p-1)r, (p-2)r, \dots, r, 0\big)$$

transforms our system (20) to a Fuchsian one with the fundamental matrix $\widetilde{Y} = \widetilde{U} z^{A'} z^E$, where in turn $\widetilde{U}(z)$ is holomorphically invertible at zero and $z^{A'} E z^{-A'}$ is holomorphic at zero. Thus,

$$Y' = z^{-C} \widetilde{U} z^{A'} z^E.$$

Consider the system with the fundamental matrix

$$Y'' = z^{(p-1)rI} Y' = z^{(p-1)rI - C} \widetilde{U} z^{A'} z^E.$$

Evidently this system has matrix of valuations equal to $A + (p-1)rI$. Since the matrices $z^{(p-1)rI-C}\widetilde{U}$ and $z^{A'} E z^{-A'}$ are holomorphic at zero one gets from Lemma 2 that

$$\operatorname{tr} A' \le \operatorname{tr}(A + (p-1)rI) = \operatorname{tr} A + (p-1)pr.$$

On the other hand

$$\operatorname{tr}\big((p-1)rI - C\big) + \operatorname{tr} A' = \operatorname{tr} A + (p-1)pr + \operatorname{ord}_0 \det P(z).$$

Thus, $\operatorname{tr}\big((p-1)rI - C\big) \ge \operatorname{ord}_0 \det P(z)$ and since $\operatorname{tr} C = p(p-1)r/2$ we arrive at the right-hand side of the inequality (19).

Consider relation (15). Its right-hand side is holomorphic at zero, while the function $B'_0(z)$ has a pole of order r there. Thus, among the diagonal entries of $P(z)$ there must be at least one of order greater than or equal to r. And we arrive at the left-hand side inequality of the (19). □

3. Irregular singular points

Let now the point zero be irregular singular for system (1). Rewrite the system in the form

$$z\frac{dy}{dz} = A(z)y, \quad A(z) = z^{-r} \sum_{n=0}^{\infty} A_n z^n, \tag{21}$$

where the series converges and $A_0 \neq 0$.

What form a fundamental matrix of the system at zero have? To find the answer let us try to simplify the form of the system. Under the transformation

$$y = T(z)y' \tag{22}$$

system (1) goes to

$$z\frac{dy'}{dz} = B(z)y', \quad z\frac{dT}{dz} = A(z)T - TB(z). \tag{23}$$

Let the matrix A_0 be of the block diagonal form $A_0 = \operatorname{diag}(A_0^{11}, A_0^{22})$. We shall try to reduce the whole system (21) to the same block diagonal form as that of the matrix A_0 with the help of an analytic transformation of the form

$$T(z) = \begin{pmatrix} I & T_{12}(z) \\ T_{21}(z) & I \end{pmatrix}, \quad T^{jk}(z) = \sum_{n=1}^{\infty} T_n^{jk} z^n. \tag{24}$$

Let us present the matrices $A(z)$ and $B(z)$ in the same block form

$$A(z) = \begin{pmatrix} A_{11}(z) & A_{12}(z) \\ A_{21}(z) & A_{22}(z) \end{pmatrix}, \quad B(z) = \begin{pmatrix} B_{11}(z) & 0 \\ 0 & B_{22}(z) \end{pmatrix}. \tag{25}$$

Then it follows from (23) that

$$\begin{pmatrix} 0 & z\,dT_{12}/dz \\ z\,dT_{21}/dz & 0 \end{pmatrix} = \begin{pmatrix} A_{11} + A_{12}T_{21} & A_{11}T_{12} + A_{12} \\ A_{21} + A_{22}T_{21} & A_{21}T_{12} + A_{22} \end{pmatrix} - \begin{pmatrix} B_{11} & T_{12}B_{22} \\ T_{21}B_{11} & B_{22} \end{pmatrix}.$$

Thus, for unknown T_{12} one gets

$$z\frac{dT_{12}}{dz} = A_{11}T_{12} + A_{12} - T_{12}A_{21}T_{12} - T_{12}A_{22}.$$

Substituting the series

$$A_{jk}(z) = \sum_{n=0}^{\infty} A_n^{jk} z^{n-r}$$

into the latter equality one gets for every $n = 1, \dots$

$$(n-r)T_{n-r}^{12} = A_n^{12} + A_0^{11}T_n^{12} - T_n^{12}A_0^{22} + \sum_{m=1}^{n-1}(A_{n-m}^{11}T_m^{12} - T_m^{12}A_{n-m}^{22})$$
$$- \sum_{\mu=1}^{n-2} T_\mu^{12} \sum_{m=1}^{n-m-1} A_{n-m-\mu}^{21} T_m^{12},$$

$$A_0^{11}T_n^{12} - T_n^{12}A_0^{22} = f(A_1, \dots, A_n, T_1, \dots, T_{n-1}). \tag{26}$$

Equation (26) provides the possibility to find $T_1, \dots, T_n, \dots$ step by step. The following technical lemma plays the crucial role in this procedure.

LEMMA 4. *Let the matrices A and B have no common eigenvalues. Then for every matrix C the equation*

$$AX - XB = C$$

has a unique solution.

PROOF. The simple calculation (see [**L**]) shows that the above matrix equation is equivalent to the following vector equation:

$$Gx = c, \quad G = A \otimes I - I \otimes B^{\mathrm{t}}, \quad x = (x_{11}, \dots, x_{1p}, x_{21}, \dots, x_{pp})^{\mathrm{t}},$$
$$c = (c_{11}, \dots, c_{1p}, c_{21}, \dots, c_{pp})^{\mathrm{t}}.$$

Since the eigenvalues of the matrix G are the numbers $a_i - b_j$, where a_i, b_j are eigenvalues of the matrices A and B respectively, the matrix G is invertible. Thus, $x = G^{-1}c$. □

Now let the eigenvalues of the matrices A_0^{11} and A_0^{22} be distinct. Then applying Lemma 4 to equation (26) one can find step by step the matrices $T_1^{12}, \dots, T_n^{12}, \dots$. Then applying the same procedure one can find matrices $T_1^{21}, \dots, T_n^{21}, \dots$. And we arrive at the desired transformation $T(z)$.

Unfortunately, in the generic situation the obtained series $T(z)$ diverges for all $z \neq 0$. Such the transformation is called *formal* and below we shall use the symbol "$\widehat{T}$" to distinguish such series.

The two obtained systems (with coefficient matrices B_{11}, B_{22}) also will be formal. Note that $B_{jj}(0) = A_0^{jj}$. So if the matrix A_0^{jj} again has a block diagonal form with blocks having distinct eigenvalues, then one can again apply the similar procedure to reduce the corresponding system to a block diagonal form, etc.

If all eigenvalues of the matrix A_0 are distinct, then with the help of a formal analytic transformation (22) we finally arrive at a diagonal system

$$z\frac{d\tilde{y}}{dz} = \begin{pmatrix} \hat{b}^1(z) & 0 & \dots & 0 \\ 0 & \ddots & \ddots & \vdots \\ \vdots & \ddots & \ddots & 0 \\ 0 & \dots & 0 & \hat{b}^p(z) \end{pmatrix} \tilde{y},$$

where $\hat{b}^i(z) = \lambda_i z^{-r} + a_1^i z^{-r+1} + \cdots$.

Thus,

$$\ln \tilde{y}^i = -\frac{\lambda_i}{r} z^{-r} + \cdots + a_{r-1}^i z^{-1} + a_r^i \ln z - a_{r+1}^i z + \cdots,$$
$$\tilde{y}^i(z) = \hat{\gamma}^i z^{a_r^i} e^{q^i(z)},$$

where $\hat{\gamma}^i$ is a formal Taylor series with constant term 1 and $q^i(z)$ is a polynomial in $1/z$ of degree r with zero constant term.

So, we arrive at the formal solution $\widetilde{Y} = \widehat{\Gamma}(z) z^{\Lambda} e^{Q(z)}$. For the initial system one has

$$\widehat{Y}(t) = \widehat{F}(t) t^{\Lambda} e^{Q(t)}, \tag{27}$$

where z is denoted by t, $\widehat{F} = \widehat{T}\widehat{\Gamma}$ is an invertible formal Taylor series, Λ is a diagonal matrix, $Q(t)$ is a diagonal matrix with entries $q^i(t)$, where every $q^i(t)$ in turn is a polynomial in $1/t$ of the form $(-\lambda_i/r)t^{-r} + \cdots + a_{r-1}^i t^{-1}$, $\lambda_1, \dots, \lambda_p$ being the eigenvalues of the matrix A_0.

The obtained solution is just formal, but it contains very important information on the system. First it satisfies equation (21) and secondly, for every sector S with the vertex at zero and of opening smaller than π/r there exists a proper solution Y_S such that the formal series $\widehat{F}(z) = F_0 + F_1 z + \cdots$ is *asymptotic* to the function $Y_S e^{-Q(z)} z^{-\Lambda}$. The latter means that for every n

$$z^{-n}\left(Y_S e^{-Q(z)} z^{-\Lambda} - \sum_{i=0}^{n} F_i z^i\right) = o(1)$$

as z tends to zero in the sector S (see [**W**]).

If one has two sectors S_i, S_j with a nonzero intersection, then the corresponding proper solutions Y_{S_i}, Y_{S_j} from the theorem are connected as follows: $Y_{S_i} = Y_{S_j} C_{ji}$. The matrices C_{ij} are called *Stokes matrices* and they explain how exponential asymptotics of proper solutions vary from sector to sector. (For detailed information on Stokes phenomena see [**S1, S2, M, W**].)

We investigated the case of distinct eigenvalues of the leading term A_0 of equation (21). This case is called nonresonant. In the resonant case the situation becomes much more complicated. And the following statement holds (see [**W, B2**]).

THEOREM 3. *Equation* (21) *has a formal solution of the form* (27), *where* t *equals* $z^{1/s}$, s *is the least common multiple of the numbers* $2, 3, \dots, p$, $\widehat{F}$ *is an invertible formal meromorphic series in* t, $Q(t) = \operatorname{diag}(q^1(t)I_1, \dots, q^l(t)I_l)$, q^i *is a polynomial in* $1/t$ *of degree* rs *without constant term,* I_i *is the identity matrix of* *size m_i, and Λ is a block diagonal matrix with upper triangular blocks Λ^i of sizes m_i.*

For every sector S in the t-plane with the vertex at zero and of opening smaller than π/rs there exists a proper solution Y_S such that the formal series $\widehat{F}(t) = F_{-k}t^{-k} + \cdots + F_0 + F_1 t + \cdots$ is asymptotic to the function $Y_S e^{-Q(t)} t^{-\Lambda}$ in S.

Consider a neighborhood of zero in the t-plane $\mathbb{C}$ and consider the space $\widehat{X}$ generated by the columns of the matrix $\widehat{Y^0}(t) = \widehat{F}(t)t^{\Lambda}$. Due to Theorem 3

$$\widehat{X} = \widehat{X}^1 \oplus \cdots \oplus \widehat{X}^l,$$

where the subspaces $\widehat{X}^i$ correspond to the blocks Λ^i. Every subspace $\widehat{X}^i$ is invariant under the formal analytic continuation along a loop going around zero in the t-plane and every element of the subspace can be presented in the form of a finite formal logarithmic sum, which has exactly the same form (7) as a proper solution (replacing the Laurent series by a formal meromorphic series). Thus, in the same way as in the regular singular case one can determine the formal monodromy, formal valuations of elements from every subspace $\widehat{X}^i$, formal Levelt bases etc. And we arrive at the following statement.

THEOREM 4. *There exists a constant block diagonal matrix C (with blocks as Q) such that the formal Levelt fundamental matrix $\widehat{Y}_e(t) = \widehat{Y}(t)C$ of system (21) can be represented (in a neighborhood of zero in the t-plane) as follows:*

$$\widehat{Y}_e(t) = \widehat{U}(t)t^{\hat{A}}t^E e^{Q(t)}, \tag{28}$$

where $\widehat{U}(t)$ is a formal Taylor series, $\hat{A} = \operatorname{diag}(\varphi^1, \ldots, \varphi^p)$ is a diagonal integer valued matrix of formal valuations of the formal Levelt basis, E is an upper triangular matrix of the formal monodromy (in the t-plane) (with blocks as Q) and Q is the same as in (27).

The proof of the theorem repeats the proof of Theorem 1 (for a weakly Levelt basis).

DEFINITION 2. The eigenvalues ${}^t\beta^i$ of the matrix $\widehat{A} + E$ are called formal t-exponents of system (21). The numbers $\beta^i = {}^t\beta^i/s$ are called formal exponents of system (21) at zero.

As well as in the regular singular case the formal exponents are responsible for the rate of growth of formal solutions (after elimination of the exponential part) at irregular singular points. And due to the second part of Theorem 3 we get that the formal exponents are responsible for rates of growth of proper solutions (after elimination of exponential parts) at the singular point. But it turns out that from sector to sector proper solutions with the given formal exponents vary with respect to the Stokes phenomena.

Let us estimate the order of the function $\widehat{U}(t)$ from decomposition (28) at zero in the t-plane. For this purpose introduce the following notions.

The Katz index of irregularity i_K ($i_K(t)$) of system (21) at zero is the maximal degree of the polynomials $q_i(z)$ from decomposition (28) with respect to $1/z$ (to $1/t$).

The Malgrange index of irregularity irr_M ($\mathrm{irr}_M(t)$) of system (21) at zero is the sum of the degrees of all polynomials $q_1(z), \ldots, q_p(z)$ from decomposition (28) with respect to $1/z$ (to $1/t$).

Note here that if $t = z^{1/s}$, then $i_K = i_K(t)/s$, $\operatorname{irr}_M = \operatorname{irr}_M(t)/s$. Consider the system conjugate to system (21). It is a system of the same form with the coefficient matrix $-A^t$. The tensor product of system (21) and the conjugate system has the coefficient matrix $A \otimes I - I \otimes A^t$ and the formal fundamental matrix

$$\widehat{Y}_e \otimes (\widehat{Y}_e^{-1})^t = (\widehat{H} \otimes (\widehat{H}^{-1})^t)(e^{Q(t)} \otimes e^{-Q(t)}),$$

where $\widehat{Y}_e$ is from (28), $\widehat{H} = \widehat{U} t^{\hat{A}} t^E$. Thus, the Malgrange index of irregularity Irr_M for the latter system can be presented as follows:

$$\operatorname{Irr}_M = \sum_{i \neq j} m_i m_j \deg\big(q_i(z) - q_j(z)\big) = 2 \sum_{\substack{i,j=1 \\ i<j}}^{l} m_i m_j \deg\big(q_i(z) - q_j(z)\big),$$

where $\deg q(z)$ denotes the degree of the polynomial q with respect to $1/z$, m_i is the dimension of the block $q_i I_i$ from $Q(t)$ (see Theorem 3). The index $\operatorname{Irr}_M(t)$ is determined in a similar way and again one has $\operatorname{Irr}_M = \operatorname{Irr}_M(t)/s$.

The following statement was proved by E. Corel with the help of algebraic methods (see [**C2**]). We present here a simplified analytic proof.

PROPOSITION 4. *For the order of the determinant of the formal Taylor series $\widehat{U}(t)$ at zero in the t-plane the following inequalities hold*:

$$rs - i_K(t) \leq \operatorname{ord}_0 \det \widehat{U}(t) \leq \frac{p(p-1)}{2} rs - \frac{1}{2} \operatorname{Irr}_M(t), \tag{29}$$

where r is the Poincaré rank of system (21) *at zero in the z-plane.*

PROOF. We shall prove the proposition by induction with respect to the size of the system and to the Poncaré rank.

For all systems of rank zero the statement is proved, because such systems are Fuchsian at zero and $\operatorname{ord}_0 \det \widehat{U}(z) = 0$. Assume that the statement is already proved for all systems of Poincaré rank smaller than r and for all systems of Poincaré rank r and of size smaller than p. Prove the statement for a system (21) of Poincaré rank r and of size p.

For the leading term A_0 of the system two possibilities can occur:

(i) A_0 has at least two distinct eigenvalues;
(ii) all eigenvalues of A_0 are equal to λ.

Consider the first possibility. In this case the matrix A_0 can be transformed to a block diagonal form $A_0 = \operatorname{diag}(A_0^1, A_0^{22})$ such that the blocks have no common eigenvalues. Thus, applying the procedure presented at the beginning of the section one can transform the initial system to a block diagonal form with two subsystems of dimensions m and $p-m$. The formal fundamental matrices for the subsystems contain k^1 and k^2 subblocks of the blocks of the matrix $Q(t)$. Let the sizes of the corresponding subblocks be m_i^1 and m_j^2. Since the transformation is formally analytic (because it has form (24)), we need to estimate the order of the determinant of a matrix $\widehat{U} = \operatorname{diag}(\widehat{U}_1, \widehat{U}_2)$ from the formal Levelt decomposition (28) for the systems obtained.

Due to the induction assumption one has

$$
\begin{aligned}
(30)\quad & \operatorname{ord}_0 \det \widehat{U}(t) \\
&= \operatorname{ord}_0 \det \widehat{U}_1(t) + \operatorname{ord}_0 \det \widehat{U}_2(t) \\
&\leq \frac{m(m-1)rs}{2} + + \frac{(p-m)(p-m-1)rs}{2} - \frac{1}{2}\operatorname{Ind}_M^1(t) - \frac{1}{2}\operatorname{Ind}_M^2(t).
\end{aligned}
$$

Note that

$$
\begin{aligned}
\frac{1}{2}\operatorname{Ind}_M(t) - \frac{1}{2}\operatorname{Ind}_M^1(t) - \frac{1}{2}\operatorname{Ind}_M^2(t) &= \sum_{i=1}^{k^1}\sum_{j=1}^{k^2} m_i^1 m_j^2 \deg\big(q_i(t) - q_j(t)\big) \\
&\leq \sum_{i=1}^{k^1}\sum_{j=1}^{k^2} m_i^1 m_j^2 rs = m(p-m)rs.
\end{aligned}
$$

Since

$$
\frac{m(m-1)rs}{2} + \frac{(p-m)(p-m-1)rs}{2} + m(p-m)rs = \frac{p(p-1)rs}{2},
$$

from (30) one gets the right-hand side of the inequality (29).

Consider now the case ii). Due to Remark 1 one can assume that our system

$$
t\frac{dy}{dt} = \widehat{B}(t)y \tag{31}
$$

considered in the t-plane already has an upper triangular form (in the opposite case as well as in the proof of Proposition 3, from Remark 1 we can transform it to such a form with the help of a formal analytic transformation in t).

Now $Y(t)$ has form (28) with an upper triangular $\widehat{U}$ and we must estimate the order of the determinant of $\widehat{U}$ at zero.

The transformation $y' = \exp(-(\lambda/rs)t^{-rs}I)y$ transforms our system to a system (31) (where all functions are equipped with the index "prime") with the Poincaré rank rs and with a nilpotent upper triangular leading term B_{-rs}. Therefore, the shearing transformation $\widetilde{Y} = t^C Y'$, where $C = \operatorname{diag}\big((p-1),(p-2),\dots,1,0\big)$ transforms our system to a system with the fundamental matrix $\widetilde{Y} = \widetilde{U}t^{\hat{A}'}t^E e^{Q(t)}$ of Poincaré rank $rs-1$, where $\hat{A}'$ is a matrix of formal valuations of the new system and $\widetilde{U}(t)$ is a formal Taylor series. Note that these transformations do not change the Malgrange index $\operatorname{Irr}_M(t)$ of the initial system.

For the obtained system from the induction assumption one has

$$
\operatorname{ord}_0 \det \widetilde{U}(t) \leq \frac{p(p-1)}{2}(rs-1) - \frac{1}{2}\operatorname{Irr}_M(t).
$$

The last part of the proof in the case (ii) repeats the corresponding part of Proposition 3.

One has

$$
\begin{aligned}
Y' &= t^{-C}\widetilde{U}t^{\hat{A}'}t^E e^Q, \\
Y'' &= t^{(p-1)I}Y' = t^{(p-1)I-C}\widetilde{U}t^{\hat{A}'}t^E e^Q
\end{aligned}
$$

and the system with the fundamental matrix Y'' evidently has the matrix of formal valuations equal to $\hat{A} + (p-1)I$. Since the matrices $t^{(p-1)I-C}\widetilde{U}$ and $t^{\hat{A}'}Et^{-\hat{A}'}$ are formal Taylor series, one gets from Lemma 2 that

$$\operatorname{tr} \hat{A}' \leq \operatorname{tr}(\hat{A} + (p-1)I) = \operatorname{tr} \hat{A} + (p-1)p.$$

On the other hand

$$\operatorname{tr}\big((p-1)I - C\big) + \operatorname{tr} \hat{A}' + \operatorname{ord}_0 \det \widetilde{U}(t) = \operatorname{tr} \hat{A} + (p-1)p + \operatorname{ord}_0 \det \widehat{U}(z).$$

Thus,

$$\operatorname{tr}\big((p-1)I - C\big) + \operatorname{ord}_0 \det \widetilde{U}(t) \geq \operatorname{ord}_0 \det \widehat{U}(z)$$

and since $\operatorname{tr} C = p(p-1)/2$, we arrive at the right-hand side of the inequality (19).

Consider decomposition (28). One has for the initial system (31)

$$\widehat{B}(t)\widehat{U} = t\frac{d\widehat{U}}{dt} + \widehat{U}\big(\hat{A} + t^{\hat{A}}Et^{-\hat{A}}\big) + \widehat{U}t\frac{dQ}{dt}. \tag{32}$$

The right-hand side of the relation has a pole at zero of order $i_K(t)$, while the function $\widehat{B}(t)$ has a pole of order rs there. Thus, among diagonal entries of $\widehat{U}$ there must be at least one of order greater than or equal to $rs - i_K(t)$. And we arrive at the left-hand side inequality (29). □

Note here that the corresponding inequality (19) for the regular singular case becomes just a partial case of the obtained one, because for the regular case the Malgrange index and the Katz index are equal to zero and the formal valuations coincide with the proper ones determined in Section 2.

It is proved in [**BJL**] that the decomposition (28) can be represented as follows:

$$\widehat{Y}_e(z) = \widehat{H}(z)e^{Q(t)}, \tag{33}$$

where $\widehat{H}(z)$ is a formal meromorphic series in z at zero multiplied by z^{Λ} for some constant matrices Λ and S, $t = z^{1/s}$.

From Proposition 4, from the relation between formal valuations and formal t-valuations and from the latter relation we get the following statement:

LEMMA 5. *For the order of the determinant of the matrix $\widehat{H}(z)$ at zero the following inequalities hold:*

$$\operatorname{ord}_0 \det \widehat{H}(z) = \sum_{i=1}^{p} \beta_0^i + b_0,$$

where

$$r - i_K \leq b_0 \leq \frac{p(p-1)}{2}r - \frac{1}{2}\operatorname{Irr}_M. \tag{34}$$

REMARK 2. This estimate also holds for the regular singular case. It is necessary just to replace $\widehat{H}(z)$ by the matrix $Y_e(z)$ from (10), formal exponents by proper ones and to replace the indices of Malgrange and Katz by zero.

4. Fuchsian inequalities

Consider system (1). To every singular point a_i one can attach the sum of exponents $\beta_i^1, \ldots, \beta_i^p$ of the system at this point (these exponents are proper in the regular singular case and formal in the irregular case). How do we estimate the sum of the exponents of the system over all singular points? This problem is very important not only for the analytic theory of differential equations, but also for analytic number theory, for some questions of mathematical physics, etc.

It was known (see [**B1**]) that for the regular singular case the following estimate holds:

$$\sum_{i=1}^{n}\sum_{j=1}^{p}\beta_i^j \le 0$$

and this inequality becomes an equality if and only if the system is Fuchsian on the whole Riemann sphere.

Recently, E. Corel in [**C1, C2**] solved this problem using algebraic methods. We prove his result using the analytic approach. The central role in the proof is played by Propositions 3 and 4.

THEOREM 5. *For the sum of exponents of system* (1) *on the Riemann sphere the following inequality holds*:

$$-\frac{p(p-1)}{2}R + \frac{1}{2}\,\mathrm{IRR}_M \le \sum_{i=1}^{n}\sum_{j=1}^{p}\beta_i^j \le -R + \mathrm{I}_K, \tag{35}$$

where R is the sum of the Poincaré ranks of the system over all singular points, IRR_M *and* I_K *are the sums of the Malgrange indices* Irr_M *of irregularity and of the Katz indices of irregularity respectively over all singular points.*

PROOF. Consider the differential form $\operatorname{tr} B(z)\,dz$ on the Riemann sphere and the formal Levelt fundamental matrix $\widehat{Y}_i$ of our system at a_i. Due to the Liouville relation one has

$$\det \widehat{Y}_i = c_0 \exp\left(\int_{z_0}^{z} \operatorname{tr} B(z)\,dz\right)$$

in a neighborhood of a_i. So, we get

$$\operatorname{tr} B(z)\,dz = d\ln\det\widehat{Y}_i.$$

From (33) it follows that

$$\mathrm{ord}_{a_i}\det\widehat{Y}_i = \mathrm{ord}_{a_i}\det\widehat{H}(z)$$

and thus, from Lemma 5 we get

$$\mathrm{res}_{a_i}\operatorname{tr} B(z)\,dz = b_i + \sum_{j=1}^{p}\beta_i^j, \tag{36}$$

where b_i satisfies inequality (34). From the theorem on the sum of residues for the form $\operatorname{tr} B(z)\,dz$ one has

$$\sum_{i=1}^{n} b_i + \sum_{i=1}^{n}\sum_{j=1}^{p}\beta_i^j = 0, \tag{37}$$

and using inequalities (34) we arrive at inequalities (35). □

Consider a scalar linear differential equation (18) with rational coefficients and with singular points $a_1,\dots,a_n$ on the Riemann sphere. If all singular points are regular singular for the equation, then for the sum of its exponents the classical Fuchs relation obtains (see [**H, I**])

$$\sum_{i=1}^{n}\sum_{j=1}^{p}\beta_i^j = \frac{(n-2)p(p-1)}{2}. \tag{38}$$

In a generic situation, when singular points are irregular the corresponding relation was established by D. Bertrand and G. Laumon in [**B**]. We present this relation below.

How do we define formal exponents for equation (18)? Let $a_1 = 0$. Pass from the scalar equation to system (21):

$$y^1 = u, \quad y^2 = u', \quad \dots, \quad y^p = u^{(p-1)}.$$

Since a formal fundamental matrix of the constructed system has the form (27), the same is true for the first row of the matrix. But this row represents the formal basis in the space of solutions of the equation. Therefore one gets that

there exists a formal basis in the space of solutions to equation (18) *of the form* (27), *where* $\widehat{F}$ *is a formal meromorphic vector series in* t.

Thus, as in the case of a system, one can determine formal monodromy, valuations, formal Levelt basis, irregularity indices, etc. And it's clear that Theorem 4 holds for equation (18) as well, upon replacing the matrix $\widehat{U}(t)$ from (28) by a vector.

The following theorem is due to D. Bertrand and G. Laumon ([**B**]).

THEOREM 6. *For the sum of exponents of a scalar equation* (18) *with singular points* $a_1, \dots, a_n$ *on the Riemann sphere the following relation holds*:

$$\sum_{i=1}^{n}\sum_{j=1}^{p} \beta_i^j = \frac{(n-2)p(p-1)}{2} + \frac{1}{2}\,\mathrm{IRR}_M, \tag{39}$$

where IRR_M *is the sum of the Malgrange indices* Irr_M *of irregularity over all singular points.*

To prove the theorem we need the following lemma. Let $a_1 = 0$. Denote by θ the operator $zh(z)d/dz$.

LEMMA 6. *The transformation*

$$y^1 = u, \quad y^2 = \theta u, \quad \dots, \quad y^p = \theta^{p-1}u,$$

where $h(0) \neq 0$ *transforms the equation to the system with fundamental matrix of the form* (33) *such that*

$$\mathrm{ord}_0 \det \widehat{H}(z) = \sum_{j=1}^{p} \beta_1^j - \frac{1}{2}\,\mathrm{Irr}_M\,. \tag{40}$$

PROOF. First, consider the regular singular case. In this case $Q = 0$ and one can see that the valuation of every column of the fundamental matrix, constructed from a basis of the space of solutions to the equation is equal to the valuation of its first component, which is just the valuation of the corresponding solution to the equation (this follows from the fact that the order of a function y having polynomial growth at zero is bounded above by the order of growth of the function $\theta^j y$). Thus, the exponents of the constructed system coincide with that of the initial scalar equation. But the simple straightforward calculation of the coefficient matrix shows that the constructed system is Fuchsian at zero. And we arrive at the equality

$$\mathrm{ord}_0 \det \widehat{H}(z) = \sum_{j=1}^{p} \beta_1^j.$$

Consider the case in which all polynomials $q_i(t)$ from $Q(t)$ in the decomposition (27) for our equation are nonzero and distinct. In this case the leading term (the term of smallest order at zero) of the function $\det \widehat{H}(t)$ coincides with the following one:

$$w^l(t) = \begin{vmatrix} h_1 & \dots & h_p \\ h_1\theta q_1 & \dots & h_p\theta q_p \\ \vdots & \dots & \vdots \\ h_1(\theta q_1)^{p-1} & \dots & h_p(\theta q_p)^{p-1} \end{vmatrix},$$

where $(h_1, \dots, h_p)$ is the first row of the matrix $\widehat{H}(t)$. This determinant can be reduced directly to the Vandermonde determinant and we get

$$w^l(t) = \prod_{i=1}^{p} h_i \prod_{\substack{i,j=1 \\ i<j}}^{p} (\theta q_i - \theta q_j).$$

Note that

$$\operatorname{ord}_0 \prod_{i=1}^{p} h_i = \sum_{j=1}^{p} {}^t\beta_1^j$$

(the elements of the first row of the matrix $\widehat{H}(t)$ form a formal Levelt basis for the scalar equation in the t-plane). Due to the fact that

$$\operatorname{ord}_0(\theta q_i - \theta q_j) = -\deg(q_i - q_j),$$

where the degree of the polynomial is considered with the respect to $1/t$, we get in the t-plane

$$\operatorname{ord}_0 \prod_{\substack{i,j=1 \\ i<j}}^{p} (\theta q_i - \theta q_j) = -\frac{1}{2} \operatorname{Irr}_M(t),$$

$$\operatorname{ord}_0 \det \widehat{H}(t) = \sum_{j=1}^{p} {}^t\beta^j - \frac{1}{2} \operatorname{Irr}_M(t).$$

Thus, in the z-plane one gets equality (40).

In the general case (when among the polynomials q_i there are equal ones) the proof is more technical and we refer to [**B**] in this case. □

Now we have everything needed for the proof of the theorem.

PROOF OF THEOREM 6. Let the point ∞ be a point of holomorphy for equation (18). Pass from the equation to system (1) with the help of the transformation

(41) $$y^1 = u, \quad y^2 = \theta u, \quad \dots, \quad y^p = \theta^{p-1} u,$$

where $h(z) = \prod_{i=2}^{n}(z - a_i)$.

Since in a neighborhood of every singular point a_i this transformation has the form from Lemma 6, due to the lemma one has the equality (40) at every a_i.

The constructed system has the additional apparent regular singular point at infinity. Let $(e) = (e_1, \dots, e_p)$ be a basis in a space of solutions to equation (18). In the coordinate $x = 1/z$ one has

$$\theta = g(x)x^{-n+2}\frac{d}{dx}$$

with $g(0) \neq 0$. For the fundamental matrix $Y(z)$ constructed from (e) the following relation holds:

$$Y(z) = \Gamma(z)T(z)W(z), \quad \text{where } \Gamma(z) = \operatorname{diag}\left(1, \prod_{i=1}^{n}(z - a_i), \dots, \prod_{i=1}^{n}(z - a_i)^{p-1}\right),$$

$T(z)$ is holomorphically invertible at infinity and $W(z)$ is the Wronskian matrix of (e) in the coordinate z.

Since

$$\frac{d^j u}{dz^j} = \frac{1}{z^{2j}} \sum_{i=1}^{j} c_i^j z^{j-i} \frac{d^i u}{dx^i}, \tag{42}$$

one gets

$$W(z) = \Gamma_1(z)\Gamma_2(z)V(\zeta), \quad \text{where } \Gamma_1(z) = \operatorname{diag}(1, z^{-2}, \dots, z^{-2(p-1)}),$$

$\Gamma_2(z)$ is a lower triangular matrix with diagonal elements $c_j^j = (-1)^j$ and $V(x)$ is the Wronskian matrix of (e) in the coordinate x. Note that the latter matrix is holomorphically invertible at $t = 0$, because the point ∞ is a point of holomorphy for the initial equation. Thus,

$$\det Y(z) = \det\big(\Gamma(z)T(z)\Gamma_1(z)\Gamma_2(z)V(x)\big) = z^b f(z),$$

where $b = (n-2) + 2(n-2) + \dots + (p-1)(n-2) = (n-2)p(p-1)/2$ and the holomorphic function $f(z)$ does not vanish at infinity.

As in the proof of Theorem 5 consider the sum of residues of the form $\operatorname{tr} B(z)\, dz = d \ln \det Y$ for the constructed system. The residue of the form at infinity is equal to $-b = -(n-2)p(p-1)/2$, while due to (40) the sum of residues at the singular points a_i is equal to $\sum_{i=1}^{n} \sum_{j=1}^{p} \beta_i^j - \frac{1}{2}\,\mathrm{IRR}_M$. The statement of the theorem follows again from the theorem on the sum of residues. □

The Fuchs relation (for a scalar equation) (39) is valid for a compact Riemann surface M as well. Here by the term scalar equation one understands a family of such equations (one equation attached to every coordinate neighborhood) such that in each nonempty intersection of the neighborhoods the spaces of solutions to these equations coincide. One just needs to replace the term $n-2$ in (39) by $n + 2g - 2$, where $2 - 2g$ is the Euler characteristic of the surface (see again [**B**]). And one gets the following statement:

THEOREM 7. *For the sum of exponents of a scalar equation* (18) *with singular points* $a_1, \dots, a_n$ *on a compact Riemann surface* M *of genus* g *the following relation holds*:

$$\sum_{i=1}^{n} \sum_{j=1}^{p} \beta_i^j = \frac{(n + 2g - 2)p(p-1)}{2} + \frac{1}{2}\,\mathrm{IRR}_M, \tag{43}$$

where IRR_M *is the sum of the Malgrange indices* Irr_M *of irregularity over all singular points.*

PROOF. Consider a covering $\{U_i\}$ of M by small discs, such that $U_1, \dots, U_n$ have the singular points $a_1, \dots, a_n$ as centers, $U_i \cap U_j = \varnothing$, $i, j = 1, \dots, n$, $i \neq j$ and the set $U_{n+1}, \dots, U_k$ is a covering of $M \setminus \{a_1, \dots, a_n\}$.

Connect a nonsingular point z_0 with every U_i by a path γ_i with the end at z_i. Consider a basis $y_1, \dots, y_p$ in the space X of solutions to the scalar equation under consideration at z_0 and analytically continue the elements of the basis along each γ_i. We use the notation $(y_1^i, \ldots, y_p^i)$ for these continuations.

If $i > n$, then consider the matrix

$$W_i = \begin{pmatrix} y_1^i & \ldots & y_p^i \\ dy_1^i/dz & \ldots & dy_p^i/dz \\ \vdots & \ldots & \vdots \\ d^{p-1}y_1^i/dz^{p-1} & \ldots & d^{p-1}y_p^i/dz^{p-1} \end{pmatrix},$$

where z is a local coordinate in U_i.

If $i \leq n$, then instead of the matrix W_i consider the following one:

$$\widetilde{W}_i = \begin{pmatrix} y_1^i & \ldots & y_p^i \\ \theta y_1^i & \ldots & \theta y_p^i \\ \vdots & \ldots & \vdots \\ \theta^{p-1}y_1^i & \ldots & \theta^{p-1}y_p^i \end{pmatrix},$$

where $\theta = z\,d/dz$ and $z(a_i) = 0$.

Using the covering $U_{n+1}, \ldots, U_k$ and transition functions

$$g_{ij} = W_i W_j^{-1}$$

construct a bundle $\widetilde{F}$ over $M \setminus \{a_1, \ldots, a_n\}$. This bundle is equipped with the holomorphic connection $\widetilde{\nabla}$ whose form ω_i in U_i is equal to

$$\omega_i = dW_i W_i^{-1}.$$

For every $U_1, \ldots, U_n$ we also get local connections

$$\widetilde{\omega}_i = d\widetilde{W}_i \widetilde{W}_i^{-1}.$$

The local systems defined by $\widetilde{\omega}_i$ and by the restriction $(\widetilde{F}, \widetilde{\nabla})|_{U_i \setminus a_i}$ are isomorphic, so one can extend the vector bundle $(\widetilde{F}, \widetilde{\omega})$ to the bundle (F, ω) over all of M.

Let us calculate the degree of the bundle F. On the one hand due to Lemma 6

$$\deg F = \sum_{i=1}^{n} \operatorname{tr} \operatorname{res}_{a_i} \nabla = \sum_{i=1}^{n} \sum_{j=1}^{p} \beta_i^j - \frac{1}{2} IRR_M.$$

On the other hand

$$\deg F = \deg |F|,$$

but due to (42) the determinant bundle $|F|$ is equal to the $p(p-1)/2$-th tensor power of the tensor product of the following two bundles: the canonical bundle over M and the divisor bundle determined by $\{a_1, \ldots, a_n\}$. Thus,

$$\deg F = \frac{(n + 2g - 2)p(p-1)}{2}. \qquad \square$$

As for the Fuchsian inequality (35) for a linear system on a compact Riemann surface, it is exactly the same as in the case of the Riemann sphere.

Instead of a system with meromorphic coefficients on a compact Riemann surface consider a meromorphic connection ∇ on a vector bundle F over the surface. Then one can determine the Poincaré ranks of the connection at singular points, formal exponents, etc. For this purpose one just needs to consider a local holomorphic trivialization of our bundle at a singular point a_i, to write down the corresponding form ω_i of the connection, the local system $dy = \omega_i y$ of differential equations (whose solutions are horizontal sections of the connection), and then to calculate the corresponding quantities to this system. Of course, these numbers do not depend on the choice of the local holomorphic sections.

The degree of the bundle F is equal to

$$\deg F = \sum_{i=1}^{n} \operatorname{res}_{a_i} \operatorname{tr} \omega_i .$$

Due to the relation (36) instead of (37) we get

$$\deg F = \sum_{i=1}^{n} b_i + \sum_{i=1}^{n} \sum_{j=1}^{p} \beta_i^j .$$

Now using the estimates (34) as in the proof of Theorem 5, we get the following statement (see [**B3**]).

THEOREM 8. *For the sum of exponents of a connection ∇ on a vector bundle of rank p over a compact Riemann surface the following inequality holds:*

$$-\frac{p(p-1)}{2} R + \frac{1}{2} \operatorname{IRR}_M + \deg F \leq \sum_{i=1}^{n} \sum_{j=1}^{p} \beta_i^j \leq -R + \mathrm{I}_K + \deg F, \tag{44}$$

where R is the sum of the Poincaré ranks of the connection ∇ over all singular points, IRR_M and I_K are the sums of the Malgrange indices Irr_M of irregularity and of the Katz indices of irregularity respectively over all singular points.

5. Multiplicities of zeroes

Consider a system (1), a fundamental matrix $Y(z)$ of the system and an element $y_{kl}(z)$ of this matrix that is not identically zero. Denote by $\beta_i^1, \dots, \beta_i^p$ the Levelt exponents of system (1) at the point a_i (proper exponents in the regular singular case and formal exponents in the irregular case).

THEOREM 9. *The order of the zero of the function $y_{kl}(z)$ at a point z_0 satisfies the following inequality:*

$$\operatorname{ord}_{z_0} y_{kl}(z) \leq \frac{(n-2+R)p(p-1)}{2} + T, \tag{45}$$

where $R = \sum_{i=1}^n r_i$, $T = p-1$ if z_0 is a nonsingular point and $T = \max_j \operatorname{Re} \beta_i^j$ if $z_0 = a_i$ and the point a_i is regular singular (in this case, by the order of the zero we mean the real part of the order of the asymptotics of the function at this point).

PROOF. The proof represents a modification of the proof of Theorem 3.1.1 given in [**B1**]. The schema of the proof is not new (as well as the method of symmetric powers presented below), and repeats the method used in [**BB, B**].

Consider the kth row of the matrix $Y(z)$. Denote by $y_1, \dots, y_t$ the maximal set of linearly independent elements of this row that contains the element y_{kl}.

If $t < p$, we replace the fundamental matrix $Y(z)$ by the fundamental matrix $Y(z)S$ for which the kth components of the first t columns are the functions $y_1, \dots, y_t$, while the kth components of the other columns are identically zero. Denote the matrix obtained again by Y. Then, the vector space $\widetilde{X}$ generated by the last $p-t$ columns of this matrix is invariant with respect to the monodromy action.

Therefore, the exponents β_i^k, $k = t+1, \dots, p$, of this space represent a part of the set of exponents of the whole system (1) and satisfy the inequality (see [**B1, AB**]):

$$\sum_{i=1}^{n} \sum_{k=t+1}^{p} \beta_i^k \leq 0. \tag{46}$$

The vector space X generated by the functions $y_1, \dots, y_t$ is of dimension t and represents the space of solutions of a scalar equation of order t with respect to the unknown function u of the form

$$\frac{1}{w(y_1, \dots, y_t)} \begin{vmatrix} y_1 & \dots & y_t & u \\ y_1' & \dots & y_t' & u' \\ \vdots & \dots & \vdots & \vdots \\ y_1^{(t)} & \dots & y_t^{(t)} & u^{(t)} \end{vmatrix} = 0. \tag{47}$$

The singular points of this equation are the points $a_1, \dots, a_n$ and additional, so-called apparent, singular points $b_1, \dots, b_m$ (the points at which the coefficients of the equation have poles, while the solutions are holomorphic), which are the zeros of the Wronskian of the functions $y_1, \dots, y_t$.

Note that, if the order of the zero of the component $y_{kl}(z)$ (which, by construction, is among the functions $y_1, \dots, y_t$) at the point z_0 is greater than $t-1$, then this point is certainly among the singular points of the equation constructed.

On the other hand, at the points of holomorphy of the equation, its exponents are $0, 1, 2, \dots, t-1$. Denote by $\tilde{\beta}_i^j$ the exponents of equation (47) at the points $a_1, \dots, a_n$, and by ϕ_i^j the exponents at the points $b_1, \dots, b_m$. Then,

$$\phi_i^j \geq 0, \ \tilde{\beta}_i^j - \beta_i^j \geq 0, \quad i = 1, \dots, n, \ j = 1, \dots, t, \tag{48}$$

where β_i^j are the corresponding exponents of the original system (1). (The latter inequality is based on the fact that the growth "rate" of the vector solution to the system never exceeds the growth "rate" of any of its components.)

Denote by δ_i the number

$$\delta_i = \sum_{j=1}^{t} \phi_i^j - \frac{t(t-1)}{2}$$

(the "defect" of the sum of exponents at the additional singular point b_i), and by $\tilde{\delta}_i$ the nonnegative number

$$\tilde{\delta}_i = \sum_{j=1}^{t} (\tilde{\beta}_i^j - \beta_i^j).$$

According to the Fuchs relation (39),

$$\sum_{i=1}^{m} \delta_i + \sum_{i=1}^{n} \sum_{j=1}^{t} \tilde{\beta}_i^j = \frac{(n-2)t(t-1)}{2} + \frac{1}{2} \operatorname{IRR}_M. \tag{49}$$

However, according to (46), (47) and the generalized Fuchsian inequality (29), we have

$$\sum_{i=1}^{n}\sum_{j=1}^{t}\tilde{\beta}_i^j = \sum_{i=1}^{n}\tilde{\delta}_i + \sum_{i=1}^{n}\sum_{j=1}^{t}\beta_i^j \geq \sum_{i=1}^{n}\tilde{\delta}_i + \sum_{i=1}^{n}\sum_{j=1}^{p}\beta_i^j$$
$$\geq \sum_{i=1}^{n}\tilde{\delta}_i - \left(\sum_{i=1}^{n} r_i\right)\frac{p(p-1)}{2} + \frac{1}{2}\,\mathrm{IRR}_M\,.$$

Therefore, from (49) we obtain

$$\sum_{i=1}^{m}\delta_i + \sum_{i=1}^{n}\tilde{\delta}_i \leq \frac{(n-2+R)p(p-1)}{2}.$$

Now, the assertion of the theorem follows from the fact that

$$\mathrm{ord}_{z_0}\, y_{kl}(z) \leq \sum_{i=1}^{m}\delta_i + (p-1)$$

when the point z_0 is chosen among the additional singular points, and

$$\mathrm{ord}_{z_0}\, y_{kl}(z) \leq \sum_{i=1}^{n}\tilde{\delta}_i + \max_j \mathrm{Re}\,\beta_i^j$$

when z_0 is a regular singular point of the original system. □

Now, let us consider the lth tensor power of system (1). This system is of the same form but of dimension p^l and with the matrix of coefficients

$$\widetilde{B}_l(z) = B\otimes I\otimes\cdots\otimes I + I\otimes B\otimes I\otimes\cdots\otimes I + \cdots + I\otimes I\otimes\cdots\otimes I\otimes B.$$

The fundamental matrix $\widetilde{Y}_l$ of the space of solutions of this system is given by the lth tensor power of the matrix $Y(z)$. The Poincaré ranks of this system at singular points coincide with the Poincaré ranks of the original system, while the exponents represent all possible l sums of the exponents of system (1) taken with all possible repetitions.

Consider the matrix Y_l, which represents the lth symmetric tensor power of the matrix Y. This matrix has dimension

$$d(l) = \binom{p+l-1}{l}$$

and is obtained as follows. For any z, the matrix $\widetilde{Y}_l$ can be considered as the matrix of the automorphism of the space $\mathbb{C}^p\otimes\cdots\otimes\mathbb{C}^p$ that maps the subspace $\mathcal{S}$ of symmetric tensor products into itself. It is the matrix of the restriction of this automorphism into $\mathcal{S}$ (for a certain fixed basis) that serves as the matrix $Y_l(z)$. Similarly, for the matrix $\widetilde{B}_l$, one can determine its symmetrized matrix B_l.

The matrix Y_l is a fundamental matrix of the space of solutions to a system of the form (1) with the matrix of coefficients B_l having the same singular points and Poincaré ranks that are no greater than Poincaré ranks of the original system. This system is of dimension $d(l)$ and possesses the following property: for any j, all possible monomials formed by the elements of the jth column of the matrix Y constitute a certain column of the matrix Y_l

Let us apply the construction of Theorem 9 to the system with the fundamental matrix Y_l to obtain the following assertion.

THEOREM 10. *The order of zero of the monomial $q^l(z)$ formed by elements of an arbitrary column of the matrix Y at a point z_0 satisfies the following inequality:*

$$\operatorname{ord}_{z_0} q^l(z) \leq \frac{(n-2+R)d(l)(d(l)-1)}{2} + T, \tag{50}$$

where $R = \sum_{i=1}^n r_i$, $T = d(l) - 1$ if z_0 is a nonsingular point and $T = l \max_j \operatorname{Re} \beta_i^j$ if $z_0 = a_i$ is a regular singular point (in this case, by the order of the zero we mean the real part of the order of the asymptotics of the function at this point).

Consider an arbitrary homogeneous polynomial $P^l(x_1, \dots, x_p)$ of degree l. It follows from the form of the matrix Y_l that the value of this polynomial on any solution to system (1) coincides with one of the elements of the matrix $S_1 Y_l S_2$ under an appropriate choice of nondegenerate constant matrices S_1 and S_2. Since a system with the fundamental matrix $S_1 Y_l S_2$ has the same exponents and Poincaré ranks as the system with the matrix Y_l, we obtain the following proposition.

PROPOSITION 5. *Let $P^l(x_1, \dots, x_p)$ be a homogeneous polynomial of degree l. The order of the zero of P^l, considered on the trajectory of a solution $y = (y^1, \dots, y^p)$ to system (1), satisfies the following inequality (assuming that $P^l(y^1(z), \dots, y^p(z))$ is not identically zero):*

$$\operatorname{ord}_{z_0} P^l\big(y^1(z), \dots, y^p(z)\big) \leq \frac{(n-2+R)d(l)(d(l)-1)}{2} + T, \tag{51}$$

where $R = \sum_{i=1}^n r_i$, $T = d(l) - 1$ if z_0 is a nonsingular point and $T = l \max_j \operatorname{Re} \beta_i^j$ if $z_0 = a_i$ is a regular singular point (in this case, by the order of the zero we mean the real part of the order of the asymptotics of the function at this point).

Now, let $P(z, x_1, \dots, x_p)$ be an arbitrary polynomial of degree k in z and of degree l in $x_1, \dots, x_p$. Consider a certain solution $y = (y^1, \dots, y^p)$ to system (1). Let us estimate the multiplicity of the zero of the function $P\big(z, y^1(z), \dots, y^p(z)\big)$ at the point z_0 (assuming that this function is not identically zero). To this end, we consider a system of differential equations with the fundamental matrix Y_D that represents the direct sum of the matrices $Y_0 = 1$, Y, $Y_2, \dots, Y_l$. The dimension of this system is

$$D(l) = 1 + p + d(2) + \dots + d(l),$$

and its Poincaré ranks coincide with the Poincaré ranks of the original system (1).

In addition, consider the system of differential equations

$$\frac{dy}{dz} = \frac{1}{z} \begin{pmatrix} 0 & 0 & \dots & 0 \\ 0 & 1 & \dots & 0 \\ \vdots & \vdots & \ddots & \vdots \\ 0 & 0 & \dots & k \end{pmatrix} y \tag{52}$$

with the fundamental matrix $Z(z) = \operatorname{diag}(1, z, \dots, z^k)$. This is a Fuchsian system with two singular points 0 and ∞. Therefore, its Poincaré ranks and the sum of exponents are equal to zero.

Now, consider a system with the fundamental matrix $Y_{\text{fin}} = Z \otimes Y_D$ of dimension $(k+1)D(l)$. We can assume from the very beginning that the zero point is among the singular points of system (1); therefore, the points $a_1, \dots, a_n, \infty$ are the singular points of the latter, while the point ∞ is Fuchsian for it.

It follows from the construction of this system that, for any polynomial of the type considered, there exist nondegenerate constant matrices S_1 and S_2 such that the value of the polynomial on any column of the matrix Y coincides with one of the elements of the matrix $S_1 Y_{\mathrm{fin}} S_2$. Since the system with the fundamental matrix $S_1 Y_{\mathrm{fin}} S_2$ has the same exponents and Poincaré ranks as the system with the matrix Y_{fin}, we obtain the following statement.

THEOREM 11. *The order of the zero of the polynomial $P(z, x_1, \dots, x_p)$ of degree k in z and degree l in $x_1, \dots, x_p$, considered on the trajectory of a solution $y = (y^1, \dots, y^p)$ to system* (1), *satisfies the following inequality*:

$$(53)\quad \operatorname{ord}_{z_0} P\big(z, y^1(z), \dots, y^p(z)\big) \le \frac{(n-2+R)(k+1)D(l)\big((k+1)D(l)-1\big)}{2} + T,$$

where $R = \sum_{i=1}^n r_i$, $T = (k+1)D(l) - 1$ if z_0 is a nonsingular point and $T = \max(\max_j(\operatorname{Re}\beta_i^j, l\operatorname{Re}\beta_i^j), k)$ if $z_0 = a_i$ is a regular singular point (*in this case, by the order of the zero we mean the real part of the order of the asymptotics of the function at this point*).

Since the number $D(l)$ coincides with the number of integral points in a standard p-dimensional simplex (formed of all coordinate planes and the plane $x_1 + \dots + x_p = l$) with the edge of length l and since, for large l, this number is well approximated by the volume of the simplex, we obtain

$$D(l) \sim \frac{l^p}{p!}.$$

Therefore, for large l, formula (53) is rewritten as

$$\operatorname{ord}_{z_0} P\big(z, y^1(z), \dots, y^p(z)\big) \le C(k+1)^2 l^{2p},$$

where the constant C depends only on system (1) and is easily calculated from (53).

This estimate is worse than the analogous estimate obtained in [**N**] by other methods (the estimate of [**N**] is proportional to l^p). Among the advantages of estimate (53) are the explicitly calculated constants that depend only on the original system and the fact that the multiplicities of zeros are estimated at the regular singular points of the system (all other estimates that we know are derived for points of holomorphy).

In conclusion, note that the estimates obtained may prove to be useful for estimating the measure of algebraic independence of the components of solutions to (1) over $\mathbb{C}(z)$, as well as in studying the 16th tangential Hilbert problem on the number of limit cycles. (See the studies of Nesterenko [**N**], Bertrand and Beukers [**BB**], related work of A. Buium [**B4**] and papers of Novikov and Yakovenko [**NY, Y**], Kaloshin [**K**] for these two questions.)

References

[AB] D. V. Anosov and A. A. Bolibrukh, *The Riemann–Hilbert problem*, Aspects Math., vol. E22, Friedr. Vieweg & Sohn, Braunschweig, 1994.

[AI] V. I. Arnol′d and Yu. S. Il′yashenko, *Ordinary differential equations*, Dynamical Systems. I (D. V. Anosov and V. I. Arnol′d, eds.), Encyclopaedia Math. Sci., vol. 1, Springer-Verlag, Berlin, 1988, pp. 1–148.

[BV] D. G Babbitt and V. S. Varadarajan, *Local moduli for meromorphic differential equations* **169/170** (1989).

[B1] W. Balser, *From divergent power series to analytic functions*, Lecture Notes in Math., vol. 1582, Springer-Verlag, Berlin, 1994.

[B2] ———, *Formal power series and linear systems of meromorphic ordinary differential equations*, Universitext, Springer-Verlag, New York, 2000.

[BJL] W. Balser, W. B. Jurkat, and D. A. Lutz, *A general theory of invariants for meromorphic differential equations. I: Formal invariants*, Funkcial. Ekvac. **22** (1979), no. 2, 197–221.

[B] D. Bertrand, *Exposants des systèmes différentiels, vecteurs cycliques et majorations de multiplicités*, with an appendix by D. Bertrand and G. Laumon, Équations différentielles dans le champ complexe, Vol. I (Strasbourg, 1985), Publ. Inst. Rech. Math. Av., Univ. Louis Pasteur, Strasbourg, 1988, pp. 61–85.

[BB] D. Bertrand and F. Beukers, *Équations différentielles linéaires et majorations de multiplicités*, Ann. Sci. École Norm. Sup. (4) **18** (1985), no. 1, 181–192.

[B1] A. A. Bolibrukh, *Hilbert's twenty-first problem for linear Fuchsian systems*, Trudy Mat. Inst. Steklov. **206** (1994) (Russian); English transl. in Proc. Steklov Inst. Math. **1995**, no. 5.

[B2] ———, *The Fuchs inequality on a compact Kähler manifold*, Dokl. Akad. Nauk **380** (2001), no. 4, 448–451 (Russian); English transl., Proc. Russian Acad. of Sci. Ser. Math. **380** (2001), no. 4, 1–4.

[B3] ———, *Multiplicity of the components of solutions to a system with regular singular points*, Proc. Steklov Inst. Math. **236** (2002), 61–65.

[B4] A. Buium, *Finiteness results in differential algebraic geometry and Diophantine geometry.* in this volume.

[CL] E. A. Coddington and N. Levinson, *Theory of ordinary differential equations*, McGraw-Hill, New York–Toronto–London, 1955.

[C1] E. Corel, *Inégalités de Fuchs pour les systèmes differéntiels réguliers*, C. R. Acad. Sci. Paris Sér. I Math. **328** (1999), no. 11, 983–986.

[C2] ———, *Relations de Fuchs pour les systèmes differéntiels irréguliers*, C. R. Acad. Sci. Paris Sér. I Math. **333** (2001), no. 4, 297–300.

[D] P. Deligne, *Équations différentielles à points singuliers réguliers*, Lecture Notes in Math., vol. 163, Springer-Verlag, Berlin–New York, 1970.

[G] F. R. Gantmacher, *Theory of matrices,* Vols. 1 and 2, Chelsea Publishing Co., New York, 1959.

[H] P. Hartman, *Ordinary differential equations*, John Wiley & Sons, Inc., New York–London–Sydney, 1964.

[I] E. L. Ince, *Ordinary differential equations*, Dover Publications, New York, 1956.

[K] V. Kaloshin, *Around the Hilbert* 16*th problem and an estimate for cyclicity of elementary polycycles.* in this volume.

[L] P. Lancaster, *Theory of matrices*, Academic Press, New York–London, 1969.

[LD] J. A. Lappo-Danilevsky, *Mémoires sur la théorie des systèmes des équations différentielles linéaires*, Chelsea Publishing Co., New York, 1953.

[L] A. H. M. Levelt, *Hypergeometric functions.* I, Nederl. Akad. Wetensch. Proc. Ser. A **64** (1961), 361–372; II, 373–385; III, 386–396; IV, 397–403.

[LR] M. Loday-Richaud, *Solutions formelles des systèmes différentiels linéaires méromorphes et sommation*, Exposition. Math. **13** (1995), no. 2-3, 116–162.

[M] B. Malgrange, *Connexions méromorphes.* II: *Le réseau canonique*, Invent. Math. **124** (1996), no. 1-3, 367–387.

[MR] B. Malgrange and J.-P. Ramis, *Fonctions multisommables*, Ann. Inst. Fourier (Grenoble) **42** (1992), no. 1-2, 353–368.

[N] Yu. V. Nesterenko, *Estimates for the number of zeros for functions from certain classes*, Acta Arith. **53** (1989), no. 1, 29–46.

[NY] D. Novikov and S. Yakovenko, *Tangential Hilbert problem for perturbations of hyperelliptic Hamiltonian systems*, Electron. Res. Announc. Amer. Math. Soc. **5** (1999), 55–65. electronic.

[R] J.-P. Ramis, *Les séries k-sommables et leurs applications*, Analysis, Microlocal Calculus and Relativistic Quantum Theory (Les Houches, 1979), Lecture Notes in Phys., vol. 126, Springer, Berlin–New York, 1980, pp. 178–199.

[S1] Y. Sibuya, *Stokes phenomena*, Bull. Amer. Math. Soc. **83** (1977), no. 5, 1075–1077.

[S2] ———, *Linear differential equations in the complex domain: problems of analytic continuation*, Transl. Math. Monogr., vol. 82, Amer. Math. Soc., Providence, RI, 1990.

[W] W. Wasow, *Asymptotic expansions for ordinary differential equations*, Pure Appl. Math., vol. 14, Wiley-Interscience, New York – London – Sydney, 1965.

[Y] S. Yakovenko, *Quantitative theory of ordinary differential equations and tangential Hilbert 16th problem.* in this volume.

CHAPTER 3

Quantitative Theory of Ordinary Differential Equations and the Tangential Hilbert 16th Problem

SERGEI YAKOVENKO

ABSTRACT. These highly informal lecture notes aim at introducing and explaining several closely related problems on zeros of analytic functions defined by ordinary differential equations and systems of such equations. The main incentive for this study was its potential application to the tangential Hilbert 16th problem on zeros of complete Abelian integrals.

The exposition consists mostly of examples illustrating various phenomena related to this problem. Sometimes these examples give an insight concerning the proofs, though the complete exposition of the latter is mostly relegated to separate expositions.

1. Hilbert 16th problem: Limit cycles, cyclicity, Abelian integrals

In the first section we discuss several possible relaxed formulations of the Hilbert 16th problem on limit cycles of vector fields and related finiteness questions from analytic function theory.

1.1. Zeros of analytic functions. The introductory section presents several possible formulations of the question about the number of zeros of a function of one variable. All functions below are either real or complex analytic in their domains, possibly exhibiting singularities on their boundaries. We would like to stress that only *isolated* zeros of such functions are counted, so that by definition a function *identically vanishing* on an open set *has no isolated zeros* there.

The exposition goes mostly by examples that are separated from each other by the symbol ◄. A few demonstrations terminate by the usual symbol □.

1.1.1. *Nonaccumulation and individual finiteness.* A function $f(t)$ real analytic on a finite open interval $(a, b) \subset \mathbb{R}$ may have an infinite number of *isolated* zeros on this interval only if they accumulate to the boundary points a, b of the latter. Thus the *finiteness problem* of decision whether or not the given function f has only finitely many zeros in its domain reduced to studying the boundary behavior of f.

In particular, if f is analytic also at the boundary points a, b, then accumulation of infinitely many zeros to these points is impossible and hence f has only finitely many roots on the interval (a, b). However, this strong condition of analyticity can be relaxed very considerably.

EXAMPLE 1.1. Assume that $f(t)$ defined on $(0,1]$ admits a *nontrivial* asymptotic expansion of the form

$$f(t) \sim \sum_{k=0}^{\infty} c_k t^{r_k} \quad \text{as } t \to 0^+, \tag{1}$$
$$c_k, r_k \in \mathbb{R}, \quad r_0 < r_1 < \cdots < r_k < \cdots ,$$

i.e., the difference between f and a partial sum decreases as $t \to 0^+$ faster than the next remaining term (the nontriviality means that not all c_k are zero). Then in a sufficiently small semi-neighborhood of $t = 0$ the function f has the same sign as the first nonzero coefficient c_k (since the functions t^r are nonvanishing on $\mathbb{R}_+$), and hence roots of f cannot accumulate to 0, implying that f has only finitely many zeros on $(0,1]$. ◀

This example can be generalized for functions admitting asymptotic expansion in any system of *mutually comparable* functions $f_1, f_2, \ldots$, when $f_{k+1} = o(f_k)$, provided that all of them keep constant sign near the boundary point(s).

EXAMPLE 1.2. Any function on $(0,a]$, $a > 0$, representable as a finite sum of the form

$$\sum_{k,r} h_{kr}(t)\, t^r \ln^{k-1} t, \qquad r \in \mathbb{R}, k \in \mathbb{N},$$

with the functions $h_{kr}(t)$ real analytic on $[0,a]$ (i.e., including the boundary $t = 0$), cannot have infinitely many roots accumulating to $t = 0$. Indeed, the above sum expands in the monomials $t^r \ln^k t$, $r \in \mathbb{R}$, $k \in \mathbb{Z}$, that are naturally lexicographically ordered by their growth rates as $t \to 0^+$. Clearly, a function admitting an asymptotic expansion involving terms of such a form also possesses the finiteness property.

In this example we do not exclude the cases when the expansion is trivial. However, the *convergence* assumption on h_{kr} implies than in such cases the function is identically zero and has no isolated roots at all. ◀

1.1.2. *Parametric families of analytic functions, localization and cyclicity.* Consider a function $f = f(t,\lambda) = f(t,\lambda_1,\ldots,\lambda_n)$ real analytic in an open domain U of the space $\mathbb{R} \times \mathbb{R}^n$ (this means that f can be expanded in a convergent Taylor series centered around any point in U). The function f can be considered as an analytic family of functions $f_\lambda = f(\cdot,\lambda)$ defined in the domains $U_\lambda = U \cap (\mathbb{R} \times \{\lambda\})$; the latter in turn can be considered as variable subsets of the real line $\mathbb{R}$ depending on the parameter λ.

Our nearest goal is to formulate the parametric finiteness property and establish simple sufficient conditions for it, similar to the nonparametric case.

DEFINITION 1. Let A be a point set in $\mathbb{R}$ or $\mathbb{C}$. Everywhere below we denote by $\#A$ the number of *isolated* points (finite or not) of A.

DEFINITION 2. We say that the analytic family $f = \{f_\lambda\} \colon U \to \mathbb{R}$ possesses the *uniform finiteness property* if the number of isolated zeros of all functions f_λ in their respective domains U_λ is uniformly bounded by a constant independent of λ.

It would be desirable to derive the uniform finiteness from some local properties of the family f.

DEFINITION 3. The *cyclicity* of the family f at a point $(t_*, \lambda_*) \in \mathbb{R} \times \mathbb{R}^n$ is the limit (finite or not)

$$\mathcal{N}_f(t_*, \lambda_*) = \lim_{\varepsilon \to 0^+} \sup_{\|\lambda - \lambda_*\| < \varepsilon} \#\{t \in U_\lambda : |t - t_*| < \varepsilon, f_\lambda(t) = 0\} \tag{2}$$

(note that dependence on ε is monotonic). The term comes from bifurcation theory (see Section 1.2.3 below).

We stress again that the notation $\#\{\dots\}$ above means the number of isolated roots of f_λ; the upper limit may be infinite even if all these numbers are finite.

If $f_{\lambda_*} \not\equiv 0$, then cyclicity of any family containing this function can be majorized in terms of only the function itself.

EXAMPLE 1.3. Suppose that f is analytic at an interior point $(0, 0) \in U$, and $f_0 \not\equiv 0$. Then the cyclicity of f at the origin is finite. Moreover, if $t = 0$ is an isolated root of *multiplicity* μ, that is, $f(t) = ct^\mu + \cdots$, $c \neq 0$, then the cyclicity is no greater than μ, that is, $\mathcal{N}_f(0, 0) \leqslant \mu$.

This follows from the fact that the μth derivative of f is nonvanishing at the origin and hence at all sufficiently close points. But a function whose derivative $f^{(\mu)}(t)$ has a constant sign on an interval cannot have more than μ isolated zeros (even counted with multiplicities), as follows from the Rolle theorem applied μ times.

This example can be easily generalized for the case of *complex analytic functions* defined in a domain $U \subset \mathbb{C} \times \mathbb{C}^n$. In the above assumption $f_0^{(\mu)}(0) \neq 0$, the cyclicity of the family will be *exactly* μ. To prove this, one can choose a small circle around the origin and apply the Rouché theorem to it. Another possibility would be to use the Weierstrass preparation theorem. ◀

EXAMPLE 1.4. If $\dim \lambda = 1$, then any function analytic in (t, λ) near $t = t_*$, $\lambda_* = 0$ can be expanded as $f(t, \lambda) = f_0(t) + \lambda f_1(t) + \lambda^2 f_2(t) + \cdots$. If $f(t, \lambda) \not\equiv 0$, then for some finite k necessarily $f_k(t) \not\equiv 0$, and after division by λ^k the question about *isolated* zeros of f can be reduced to the situation when $f(t, 0) \not\equiv 0$, discussed earlier. In this case $\mathcal{N}_f(t_*, 0)$ is no greater than the multiplicity of f_k at t_*, where f_k is the first nonzero term in the expansion. ◀

This example illustrates an absolutely general fact about analytic functions (either real or complex): the cyclicity $\mathcal{N}_f(t, \lambda)$ takes finite values at all *interior* points of the domain of analyticity U.

THEOREM 1. *If f is analytic at a point $(t, \lambda) \in \mathbb{R} \times \mathbb{R}^n$, then $\mathcal{N}_f(t, \lambda) < +\infty$.*

This assertion can be derived from general finiteness properties of analytic sets, see [**Ł**]. In the present form the theorem was formulated in connection with bifurcations of limit cycles, see [**R3**].

The proof of finiteness of $\mathcal{N}_f$ at points of analyticity of f in the general multi-parametric case requires some analytic techniques. One possibility — assuming for simplicity the point to be at the origin $(0, 0)$ — is to consider the expansion $f(t, \lambda) = \sum_{k \geqslant 0} a_k(\lambda) t^k$ and the ideal in the ring $\mathfrak{R}$ of analytic germs at $(\mathbb{R}^n, 0)$ generated by the coefficients $a_k(\lambda)$. This ideal is called the *Bautin ideal* [**R3**]. Since the ring $\mathfrak{R}$ is Noetherian, the Bautin ideal is in fact generated by a finite number of the coefficients a_k. If the germs $a_1, \dots, a_\nu$ generate the Bautin ideal for some finite ν, then one can show that the function $f(t, \lambda)$ can be presented as $\sum_{k=0}^{\nu} a_k(\lambda)t^k h_k(t,\lambda)$ with h_k analytic in t, λ and $h_k(0,0) \neq 0$, see [**R3**]. From this it is already easy to derive that $\mathcal{N}_f(0,0) \leqslant \nu < \infty$. Almost no modification is required to cover the complex analytic case as well: the only difference is that the usual Rolle theorem cannot be applied to holomorphic nonreal functions. The alternative is to use the complex Rolle theorem from [**KY**]. An example of such use in a situation very similar to that discussed above can be found in [**Y3**].

1.1.3. *Finite cyclicity and uniform finiteness.* The main reason for introducing the notion of cyclicity is the following very simple but basic theorem. In applications to bifurcation of limit cycles of analytic vector fields it was observed by R. Roussarie [**R1, R2, R3**].

Let $f\colon U \to \mathbb{R}$ as before be an analytic family of functions and $\mathcal{N}_f\colon \overline{U} \to \mathbb{N} \cup \{+\infty\}$ its cyclicity function.

THEOREM 2. *If the closure $\overline{U}$ is a compact subset of $\mathbb{R} \times \mathbb{R}^n$, and the cyclicity $\mathcal{N}_f$ is finite everywhere on $\overline{U}$, then the family $f = \{f_\lambda\}$ admits a uniform upper bound on the number of isolated roots:*

$$\sup_{\lambda} \#\{t \in U_\lambda \colon f_\lambda(t) = 0\} < +\infty. \tag{3}$$

PROOF. By definition of the counting function, any point $(t_*, \lambda_*) \in \overline{U}$ can be covered by a sufficiently small cube $\{|t - t_*| < \varepsilon, \|\lambda - \lambda_*\| < \varepsilon\}$ such that the number of isolated roots of f in this cube is no greater than the number $\mathcal{N}_f(t_*, \lambda_*)$, finite by the assumptions of the theorem. It remains only to choose a finite subcover of the compact $\overline{U}$ and add together the corresponding cyclicities. □

Theorem 1 asserts that the cyclicity is automatically finite at all *interior* points of the domain U, so it is again the boundary behavior of a given parametric family that determines whether this family possesses the uniform finiteness or not.

1.1.4. *Terminology: individual vs. existential finiteness problems.* The arguments proving finiteness of the number of isolated zeros of an individual analytic function $f(t)$ and the arguments establishing uniform boundedness of this number for parametric families of such functions are both of the same purely existential nature: neither of them gives any way to *compute* or even *estimate from above* these numbers. However, the parametric claim is definitely stronger than the assertion concerning individual functions of this family (one can easily construct families in which the number of isolated zeros is always finite but not uniformly bounded).

We shall repeatedly encounter the problem on bounding the number of isolated zeros for various classes of analytic functions and finite parameter families of such functions, mostly defined by ordinary differential equations with polynomial right-hand sides. For each class one can pose several finiteness-type problems in increasing order of strength (the gap in the enumeration will be explained below).

1. *Individual finiteness problem.* Prove that each function from the family possesses only finitely many isolated zeros.

2. *Existential (uniform) finiteness problem.* Prove that the number of isolated roots is uniformly bounded over all functions from this family.

4. *Constructive finiteness problem.* Find an explicit upper bound for the number of isolated roots or at least find an algorithm for computing this bound.

The adjective "existential" stresses the difference between the last two types of problems, whereas the adjective "uniform" will be used to underscore the difference between the first two assertions.

EXAMPLE 1.5. Consider the class of parametric families of functions real analytic on $[0,1]$ depending analytically on parameters from $[0,1]^n \subset \mathbb{R}^{n+1}$ (in both cases including the boundary). As was just explained, the individual finiteness for functions from this class follows from the uniqueness theorem for analytic functions while the existential (uniform) finiteness theorem (Theorem 2) follows from Theorem 1. ◄

EXAMPLE 1.6. For any finite d, polynomials in one variable of degree $\leqslant d$ form a finite-parameter family. The constructive finiteness theorem for this "toy" class of functions is known as the Fundamental Theorem of Algebra. Less trivial examples can be found in Section 3.1. ◄

1.1.5. *Constructive finiteness.* In practice the existential finiteness is always derived from finite cyclicity using Theorem 2. Both steps (finite cyclicity and its globalization) use arguments of an existential nature. However, at least theoretically it may happen that the function $\mathcal{N}_f$ is explicitly bounded at all points of the closure $\overline{U}$. This still would not allow one to compute explicitly the global uniform bound on the number of isolated zeros of f, but such *local constructive finiteness* would clearly be a much stronger assertion concerning the family f. The corresponding finiteness problem occupies an intermediate place between existential and (global) constructive finiteness problems.

3. *Constructive finite cyclicity problem.* Find an explicit majorant for the counting function $\mathcal{N}_f$.

Yet it should be remarked that in order to discuss the constructive finiteness problems (global or local), the family should be defined by some algebraic data, otherwise computability does not make sense. We postpone discussion of these questions until Section 3.1.

EXAMPLE 1.7. Let $F_1(t), \dots, F_n(t)$ be analytic functions (real or complex) defined in a domain U, satisfying together a system of polynomial ordinary differential equations

$$\frac{dF_i}{dt} = \sum_{k+|\alpha|\leqslant d} c_{ik\alpha}\, t^k F^\alpha, \qquad F^\alpha = F_1^{\alpha_1} \cdots F_n^{\alpha_n},\ i = 1, \dots, n. \tag{4}$$

Assume the degree d and all coefficients $c_{ik\alpha}$ (real or complex) known. Then for any finite $m \in \mathbb{N}$ the polynomial combinations of degree $\leq m$ of these functions constitute a finite-parameter family $\mathcal{F}_m$: the coefficients $\lambda = \{\lambda_{k\alpha} : k + |\alpha| \leqslant m\}$ of the polynomial combinations $f(t,\lambda) = \sum_{k+|\alpha|\leqslant m} \lambda_{k\alpha}\, t^k F^\alpha(t)$ are the parameters. The domain of this family is clearly the Cartesian product of the linear space of the appropriate dimension and the common domain U of the functions F_i. Though solutions of polynomial systems may blow up in finite time, we will assume for simplicity that all F_i are analytic on the compact closure $\overline{U}$ of U.

In these assumptions both individual and existential finiteness are guaranteed by Theorem 2 which, as usual, does not provide any bound on the number of zeros. Yet, using algebraicity of the system defining these functions, one can derive explicit bounds. In [**G1**] A. Gabrielov found an upper bound for the maximal *order of zero* (*multiplicity of a root*) that a nontrivial (not identically equal to zero) function f from the family $\mathcal{F}_m$ may have, in terms of n, d and m. The answer is given by an expression polynomial in the degrees d, m and exponential in n (the dimension of the system). Y. Yomdin in [**Y4**] derived from this multiplicity bound the corresponding *cyclicity bound*, using arguments involving the Bautin ideal and its generators (a subtle generalization of the approach outlined in the discussion of Theorem 1). This combination of the two results gives a complete solution for the constructive finite cyclicity problem for polynomial combinations of functions defined by systems of polynomial ordinary differential equations. ◀

The corresponding global problem was settled by D. Novikov and the author [**NY2, NY4**]. Unlike the bound for cyclicity, the bound for the total number of zeros depends on the size of the domain $\rho = \max_{t\in U} |t|$ where they are counted and, rather naturally, on the magnitude of the coefficients $R = \max_{i,k,\alpha} |c_{ik\alpha}|$ of the system (4). The answer is polynomial in ρ, R but as a function of the other (integer) parameters n, m, d is a tower of height 4. For height of a tower see Example 3.6. The detailed explanation of this result is postponed until Section 4.1.

1.1.6. *Paradigm.* All the above results on elementary properties of analytic functions and their parametric families were mentioned in order to set out a paradigm that will be used when discussing much more complicated parallel problems on limit cycles of analytic and polynomial vector fields. In order to make the similarity transparent, all finiteness theorems can be organized in the form of a table with finiteness types labelling the rows and corresponding theories represented by different columns. The first column describes the theory of analytic functions as in Example 1.5 above. Cells of this column list the key theorems ensuring each finiteness type.

The following section Section 1.2 describes some known results fitting the second column of Table 1: the strongest form of the Hilbert 16th problem should occupy the place where the question marks appear. The parallel nonconstructive results for Abelian integrals will be discussed in Section 1.4: their complexity level can be described as intermediate between relatively elementary theorems on analytic functions and transcendentally difficult counterparts on limit cycles.

REMARK. In general, the results in the lower part of the table are more difficult than those from the upper part. However, this is not true when comparing existential finiteness with constructive finite cyclicity: while the former deals with boundary behavior, the latter addresses issues of a completely different nature that can be (and indeed sometimes are) simpler. In particular, this is the case with Abelian integrals, where the bound for multiplicity can be relatively easily obtained, see [**M**] and Section 2.2.5 below.

1.2. Limit cycles of planar vector fields. This section briefly surveys a few known general finiteness results for limit cycles of vector fields.

1.2.1. *Basic facts.* A system of two polynomial differential equations

$$\dot{x} = P(x,y),\ \dot{y} = Q(x,y), \quad P, Q \in \mathbb{R}[x,y] \tag{5}$$

defines a vector field on the real plane $\mathbb{R}^2$, spanning the distribution (line field) with singularities given by null spaces $\{\Omega = 0\}$ of the polynomial Pfaffian form $\Omega = Q\,dx - P\,dy$. If P and Q have a nontrivial greatest common divisor R, then this distribution on the Zariski open set $\{R \neq 0\}$ coincides with that given by

TABLE 1. Various flavors of the Hilbert 16th problem

Finiteness type	Category of objects		
	Functions analytic on a compact	*Limit cycles of polynomial vector fields*	*Abelian integrals*
Individual finiteness	uniqueness theorem	Ilyashenko (1991) – Écalle (1992)	easy exercise see §1.4.1
Existential finiteness	finite cyclicity Theorems 1 and 2, see §1.1.3	partially known for quadratic case (about 90 out of 121 polycycles listed in [**DRR**])	Varchenko – Khovanskiĭ (1985)
Constructive finite cyclicity	Gabrielov – Yomdin (1998)	known for generic elementary (Ilyashenko – Yakovenko, 1991) and few other polycycles	Mardešić [**M**] (multiplicity)
Constructive finiteness	Novikov – Yakovenko (1999)	??? (nothing in view)	see Table 2

the form $R^{-1}\Omega$ and hence extends analytically to all but finitely many points of the curve $\{R = 0\}$. For the same reasons the distribution extends to the infinite line $\mathbb{R}P^1$ on the projective plane $\mathbb{R}P^2$, having at most finitely many isolated singularities on the latter after cancellation of an eventual common factor. In other words, when dealing with an individual system of differential equations (5), one can assume that the corresponding distribution has only *isolated singularities.* It is sometimes convenient to talk about *foliations with isolated singularities* defined by the distribution $\{\Omega = 0\}$ on $\mathbb{R}P^2$.

A limit cycle γ by definition is an isolated periodic trajectory of (5). Limit cycles of a vector field can be tracked by their intersections with analytic arcs transversal to the field. More precisely, near any intersection $p = \gamma \cap \sigma$ of a periodic orbit γ with a transversal arc σ, the Poincaré first return map Δ_σ, the transport along integral curves until the next intersection with the transversal, is defined. It has p as a fixed point, *isolated* if γ is a limit cycle. Choosing an analytic chart t on σ allows one to describe p as a root of the *displacement function* $f(t) = \Delta_\sigma(t) - t$. All nearby limit cycles must intersect σ in points that are isolated roots of the displacement f.

By the general theorem on analytic dependence of solutions of differential equations on initial conditions and parameters, the displacement function f is analytic near p. As a corollary, we conclude that an infinite number of limit cycles (corresponding to isolated roots of the displacement function) cannot accumulate to a periodic orbit of the field. In a similar way, if the vector field depends analytically on the parameters λ and exhibits a periodic orbit γ (isolated or not) for one value λ_* of the parameters, then there exists a finite upper bound for the number of limit cycles in a small annulus around γ, uniform over all values of λ sufficiently close to λ_*.

In general, one transversal cannot "serve" all limit cycles, and there is no natural way to define the maximal domain of the Poincaré return map unambiguously and globally. However, the discussion in Section 1.1 suggests that it is the boundary behavior that is important for counting isolated zeros of the displacement.

1.2.2. *Polycycles and limit periodic sets.* Instead of trying to reduce formally the global investigation of limit cycles to that of return maps for one or several transversal arcs, it is better to study the cycles (compact leaves of the foliation) themselves.

For the nonparametric case, the analogue of a boundary point of the domain of the Poincaré return map is a *polycycle*, an invariant set consisting of one or more *singular points* of the vector field, and a number of bi-infinite trajectories connecting them in a cyclic order (repetitions of singular points allowed). More accurately, one can show that the only sets that can appear as Hausdorff limits of periodic orbits of a vector field having only isolated singular points are polycycles.

In the parametric case one cannot assume any longer that singularities of the foliation are isolated (the polynomials P and Q may have common factors for some values of the parameters and be mutually prime for the rest). Still one can show [**FP**] that the Hausdorff limit of a family of limit cycles occurring for converging values of the parameters must be a *limit periodic set* (also known as *graphic*), an object differing from a polycycle in only one instance, namely, it may contain analytic arcs of nonisolated singularities.[1]

1.2.3. *Individual finiteness theorem and finite cyclicity.* Following the paradigm set out in Section 1.1.6, one can easily formulate counterparts of the individual and existential finiteness problems for limit cycles for vector fields. The individual finiteness theorem (known also as the *Dulac conjecture* after being believed for some 60 years to be the *Dulac theorem*) asserts that *any polynomial vector field can have only finitely many limit cycles.* This is an easy corollary to the highly nontrivial *nonaccumulation theorem* asserting that limit cycles of an analytic vector field cannot accumulate to any polycycle. The latter assertion is the most spectacular and the most general fact established so far in connection with the Hilbert 16th problem, which was independently proved by Ilyashenko [**I4**] and Écalle [**É**] by totally different methods. This is a typical example of an assertion on boundary behavior of the Poincaré map.

REMARK. It would be appropriate to notice here that Dulac reduced the individual finiteness problem to the situation discussed in Example 1.1. However, he did not notice that the expansion (1) (only involving logarithms) may well be *trivial*, hence not allowing for so easy a treatment. It took both Ilyashenko and Écalle hundreds of pages to prove that even in this apparently highly degenerate case the nonaccumulation still holds.

The existential finiteness problem for polynomial vector fields arises very naturally, since coefficients of the polynomials P, Q can be treated as parameters. These parameters can be considered as ranging over the compact projective space of a suitable dimension, since simultaneous multiplication of both P and Q by a nonzero constant does not alter the phase portrait. Existence of a uniform upper bound for the number of limit cycles would mean that *for any degree d the number*

[1] The precise definition of graphics and polycycles varies slightly from one source to another.

of limit cycles for a polynomial vector field of degree $\leqslant d$ is bounded by some number $\mathcal{H}(d)$ depending only on d. According to the Roussarie localization theorem (an analog of Theorem 2 for limit cycles, see [**R1**]), it would be sufficient to prove that any limit periodic set has finite cyclicity within the universal family of polynomial vector fields of degree $\leqslant d$.

However, even for the simplest nontrivial case of quadratic ($d = 2$) vector fields, this is not proved. Dumortier, Roussarie and Rousseau composed in [**DRR**] a list of 121 graphics occurring for quadratic vector fields, and reduced the existential finiteness problem to proving that all these graphics have finite cyclicity. The first several cases were studied in the same paper by applying relatively standard tools of bifurcation theory. Since then many other considerably more delicate cases were investigated and their finite cyclicity proved, though the ultimate goal of proving finite cyclicity in the quadratic case is not yet achieved. Obviously, such a type of case-by-case study is impossible for higher degree.

On this background the perspective of *computing* or even *explicitly majorizing* the "*Hilbert number*" $\mathcal{H}(d)$, which is the strongest (constructive) form of the initial Hilbert's question "on the number and position of limit cycles" [**H4**], looks very remote.

1.2.4. *Digression: constructive solutions of localized problems.* As was already noted above, though finiteness of cyclicity for all graphics is not proved, one can independently work towards obtaining constructive bounds for cyclicity in cases when its finiteness is already known. This is possible since cyclicity is very strongly dependent on the types of singular points lying on the polycycle (resp., the limit periodic set).

For example, if all singular points on the polycycle are elementary (having nonnilpotent linear parts), then cyclicity of *generic* r-parameter families can be estimated in terms of the number r of independent parameters. The corresponding algorithm was suggested by Ilyashenko and the author [**IY2**] and improved to a very concise and explicit answer by Kaloshin [**I5**]. Here the genericity means transversality of the r-parameter family to some real analytic subset in the properly defined space of jets of finite order; the family itself can even be smooth and not necessarily analytic.

Some other cases of polycycles (usually carrying one or at most two singular points) also have known cyclicity, see [**KS**] and references therein for a synopsis. These results are partially reflected in Table 1.

Yet, if the problem is formulated for polynomial vector fields and the bounds are required to be given in terms of the degree, the problem immediately becomes transcendentally difficult. Almost nothing is known, for example, on the maximal multiplicity of a limit cycle of a polynomial vector field. Cyclicity of singular points is a considerably more "algebraic" problem. For example, when the linear part of the singularity is a nondegenerate rotation, a polynomial algorithm can be suggested which stops at a step number N if and only if the cyclicity of a singular point is N. However, the running time of this algorithm is absolutely unknown (see the discussion of a similar problem in Section 1.3.3). Recently M. Briskin, J.-P. Françoise and Y. Yomdin treated in detail one problem on Abel equations that may give a clue to understanding some of the phenomena, see, e.g., [**BFY1, BFY2**].

The following section describes one very important case when the local (with respect to parameters) constructive finiteness can be reduced to an algebraic context.

1.3. Abelian integrals: appearance and basic properties. Consider a family of analytic vector fields X_λ analytically depending on parameters $\lambda \in (\mathbb{R}^n, 0)$ (we consider only a sufficiently small neighborhood of the origin in the parameter space). Assume that for $\lambda = 0$ the vector field X_0 possesses a limit cycle γ_0 of multiplicity $\mu < +\infty$. As illustrated by Example 1.3, the cyclicity of γ_0 in this case is at most μ, being thus explicitly bounded.

The problem becomes less trivial if the field X_0 possesses an annulus filled by nonisolated periodic orbits, γ_0 being one of them. Vector fields with this property are called integrable[2]: the field admits a real analytic first integral in this annulus. Such behavior, very unlikely for an arbitrary analytic family, becomes an event of finite codimension when *polynomial* vector fields are considered.

To study bifurcations of limit cycles in perturbations of integrable systems, an approach described in Example 1.4 can be used.

1.3.1. *Perturbation of polynomial vector fields.* Assume that the form $\Omega = Q\,dx - P\,dy$ defining the polynomial distribution $\{\Omega = 0\}$ is *exact*: $\Omega = dH$, where $H = H(x, y)$ is a real polynomial in two variables. The corresponding vector field is then a Hamiltonian one,

$$\dot{x} = -\frac{\partial H}{\partial y}, \quad \dot{y} = \frac{\partial H}{\partial x}. \tag{6}$$

with H being its *Hamiltonian function* or simply the *Hamiltonian.* Hamiltonian vector fields are *conservative*: their divergence is identically zero.

Since dH vanishes on the null spaces of the distribution $\{\Omega = 0\}$, each leaf of the corresponding foliation belongs to a level curve of H. In particular, a periodic orbit must be a compact oval of some level curve and hence all nearby close leaves must also be closed. Hence a Hamiltonian vector field cannot have limit cycles: the corresponding Poincaré return map is the identity and the displacement identically zero.

Consider the following *one-parameter family* of polynomial distributions, perturbing the Hamiltonian distribution:

$$dH + \varepsilon\omega = 0, \qquad \varepsilon \in (\mathbb{R}^1, 0), \ \omega = p\,dx + q\,dy, \ p, q \in \mathbb{R}[x, y]. \tag{7}$$

The polynomial 1-form ω can be arbitrary.

Consider an analytic segment σ transversal to the oval $\gamma(t_*) \subset \{H = t_*\}$. This transversality implies that the Hamiltonian H, restricted to σ, gives an analytic chart t on σ. On the other hand, for all t sufficiently close to t_*, one can unambiguously choose an oval $\gamma(t) \subset \{H = t\}$ so that $\gamma(t)$ tends to $\gamma(t_*)$ as $t \to t_*$, e.g., in the sense of Hausdorff distance (this choice in a more broad context is discussed below, in Section 2.1).

The following elementary computation gives the *first variation* of the displacement function $f(t, \varepsilon) = \Delta_\sigma(t) - t$ with respect to the small parameter ε at $\varepsilon = 0$.

---

[2]This should not be confused with explicit integrability of a vector field by quadratures.

LEMMA 1 (Poincaré–Pontryagin).

$$f(t,0) \equiv 0, \qquad \frac{d}{d\varepsilon}\bigg|_{\varepsilon=0} f(t,\varepsilon) = -\oint_{\gamma(t)} \omega. \tag{8}$$

PROOF. Consider the leaf (analytic curve) of the foliation $\{dH + \varepsilon\omega = 0\}$ passing through the point t on the transversal, and denote by $\gamma(t,\varepsilon)$ the (oriented) segment of this leaf between the initial point and the next intersection with σ. By definition of the chart t, the displacement function measured in the chart t is the difference of the values of H at the endpoints of $\gamma(t,\varepsilon)$, hence

$$f(t,\varepsilon) = \int_{\gamma(t,\varepsilon)} dH$$

(the equality is exact). The form $dH + \varepsilon\omega$ vanishes on $\gamma(t,\varepsilon)$, therefore the integral above is equal to the integral of $-\varepsilon\omega$ along $\gamma(t,\varepsilon)$ and, since $\gamma(t,\varepsilon)$ converges uniformly to the *closed* curve $\gamma(t) = \gamma(t,0) \subset \{H = t\}$ as $\varepsilon \to 0$, we conclude that

$$f(t,\varepsilon) = -\varepsilon \oint_{\gamma(t)} \omega + o(\varepsilon) \quad \text{as } \varepsilon \to 0,$$

with $o(\varepsilon)$ uniform and analytic in t and ε. This yields the formula (8) for the derivative. □

Recall that an *Abelian integral* is the integral of a polynomial 1-form along any algebraic oval (i.e., the closed oval of a real algebraic curve). Clearly, Abelian integrals can be considered as *functions of the parameters*, the coefficients of the form and the algebraic equation defining the curve. However, in applications to bifurcation theory, one coefficient is clearly distinguished and plays the role of the argument, while others are treated as "true" parameters. Making a minor abuse of language, we shall always consider Abelian integrals as analytic functions in the following sense.

DEFINITION 4. A *real* (complete) Abelian integral corresponding to a Hamiltonian H and a polynomial 1-form $\omega = p(x,y)\,dx + q(x,y)\,dy$, with $H, p, q \in \mathbb{R}[x,y]$, is a multivalued function $I(t)$ of the real variable t defined by integration of ω over a real oval $\gamma(t)$ of the algebraic curve $\{H(x,y) = t\} \subset \mathbb{R}^2$.

The reason for multivaluedness is obvious, since there can be several ovals lying on the same level curve $\{H = t\}$. In Section 2.1 we describe the branching of the Abelian integral $I(t)$ after analytic continuation into the complex domain. This complexification of Abelian integrals explains the connections between different branches of these functions, see Section 2.1.

1.3.2. *Bifurcation of limit cycles from periodic orbits of Hamiltonian systems.* Limit cycles appearing in the one-parameter system (7) correspond to isolated zeros of the displacement function $f(t,\varepsilon)$, real analytic in t and ε, whose first Taylor term was just computed in Lemma 1:

$$f(t,\varepsilon) = I_0(t) + \varepsilon I_1(t) + \varepsilon^2 I_2(t) + \cdots, \quad I_0(t) \equiv 0,\ I_1(t) = -\oint_{\gamma(t)} \omega. \tag{9}$$

This is exactly the situation treated in Example 1.4: if the first variation is nontrivial, $I_1(t) \not\equiv 0$, then the number of isolated roots of $f(t,\varepsilon)$ on any interval of analyticity (i.e., not containing singularities at the endpoints) is no greater than the number of (necessary isolated) roots of the Abelian integral I_1 defined by the polynomial H and the polynomial 1-form ω. Computing or majorizing the number of isolated zeros of Abelian integrals is the central theme of these notes: under the name *Tangential Hilbert problem* it is discussed below.

The first variation method does not work if $I_1(t) \equiv 0$. In this case higher variations $I_k(\cdot)$, $k = 2, 3, \dots$, have to be computed and analyzed.

The computation of higher variations is relatively simple. It follows from results of Ilyashenko [**I2**] and Gavrilov [**G2**] that for almost all Hamiltonians H (and certainly for Hamiltonians transversal to infinity as defined below in Section 2.2, with isolated critical points and pairwise different critical values), the condition $I_1 \equiv 0$ implies that there exist two polynomials $G, F \in \mathbb{R}[x, y]$ such that

$$\omega = G(x,y)\,dH + dF(x,y), \qquad \text{hence } d\omega = dG \wedge dH. \tag{10}$$

(the converse statement is obvious).

EXAMPLE 1.8. (see [**F2**]). Consider the Hamiltonian $H(x, y) = \frac{1}{2}(x^2 + y^2)$. In the complex coordinates $z = x + iy$, $\bar{z} = x - iy$ we have $H = \frac{1}{2}z\bar{z}$. One can easily verify that the polynomial 1-form $\omega = A(z,\bar{z})\,dz + B(z,\bar{z})\,d\bar{z}$ has identically vanishing integral over the circles $\{H = t\}$ if and only if the differential $d\omega = (-A_{\bar{z}} + B_z)\,dz \wedge d\bar{z}$ contains no monomial terms of the form $(z\bar{z})^k\,dz \wedge d\bar{z}$. Any other monomial can obviously be represented in the form (10): $z^i\bar{z}^j\,dz \wedge d\bar{z} = dG \wedge dH$ with $G = z^i\bar{z}^j/(i-j)$. ◀

Using the representation (10), the formula for the second variation of the displacement map can again be expressed as an Abelian integral.

LEMMA 2 ([**Y1, F2**]). *Assume that for the perturbation* (7) *the first variation vanishes identically, so that*

$$I_1(t) = \oint_{\gamma(t)} \omega \equiv 0,$$

and G is any polynomial satisfying the condition (10). *Then the second variation of the displacement map is given by the following Abelian integral,*

$$I_2(t) = \frac{d^2}{d\varepsilon^2}\bigg|_{\varepsilon=0} f(t,\varepsilon) = -\oint_{\gamma(t)} G\omega. \tag{11}$$

This construction can be further iterated [**F2**] as long as necessary to obtain a variation that is not identically vanishing. If, on the other hand, all integrals obtained in this recurrent process turn out to be identically zero, then the displacement function itself is identically zero and hence the perturbation (7) consists of integrable systems for all ε.

REMARK. The exact conditions on the Hamiltonian H for the identity (10) to hold for any form whose integral is identically zero involve connectedness of the complex affine level curves $H^{-1}(t)$ for almost all $t \in \mathbb{C}$ [**G2**]. Discussion of the multidimensional situation can be found in [**BD**]. As soon as these conditions fail (even because of coinciding critical values, for example), then computation of higher variations may involve integration of nonpolynomial forms, see [**I1**].

1.3.3. *Open problems: bifurcation of limit cycles from Hamiltonian polycycles and generalized Poincaré center-focus problem.* It would be wrong to think that the above approach based on computing consecutive variations *completely* reduces the question on limit cycles born by perturbation of Hamiltonian systems to investigation of Abelian integrals.

First, even assuming the simplest case $I_1 \not\equiv 0$, we cannot in general say anything about limit cycles born from *critical* level curves (corresponding to polycycles of the unperturbed Hamiltonian system). In the particular case of *separatrix loops* (parts of level curves homeomorphic to the circle and carrying only one nondegenerate saddle point), the problem was settled by R. Roussarie [**R2**]. He estimated how many derivatives of the Abelian integral $I(t)$ should have zero limits as $t \to t_*$ in order for the perturbation (7) to produce n or more limit cycles close to the separatrix loop on the critical level curve $\{H = t_*\}$. This result yields an upper bound for the cyclicity of the separatrix loop in terms of the "multiplicity of the root" of the Abelian integral I_1 at the point t_* where I_1 in fact loses its analyticity. The only other type of critical level curve that can occur for a generic polynomial H is the eight-shaped curve (also carrying a nondegenerate saddle). The answer in this case is not yet known, though one should not expect serious difficulties here.

The second, apparently much more difficult problem, appears in connection with the natural question, *How many consecutive variations I_k should be computed in order to guarantee that the perturbation is nonconservative?*

The inductive process described in Lemma 2 is algebraic in the following sense. Starting from the perturbation (7) given by the form $\omega = \omega_1$, we construct a sequence of polynomial 1-forms $\omega_2, \omega_3, \dots$ which express the higher variations $I_k(t)$ (9), assuming that all previous variations vanish identically, $I_1 \equiv \cdots \equiv I_{k-1} \equiv 0$. Coefficients of the forms ω_k are given by algebraic (polynomial) expressions involving coefficients of the initial form ω. Vanishing of their integrals implies an infinite number of polynomial identities between coefficients of the initial form ω. By the Hilbert basis theorem, all these infinitely many identities are corollaries to only finitely many of them. Thus only finitely many steps of the inductive process must be performed (their number $N < +\infty$ depends on the *degree* of the initial form ω and, naturally, on the Hamiltonian H). If all integrals $\oint \omega_k$, $k = 1, \dots, N$ are identically zero, then all higher variations are necessarily zero and the family (7) has no limit cycles at all in the area considered.

However, the problem of finding an upper bound for the number N in terms of ω and H is wide open: even for the most simple case $H = \frac{1}{2}(x^2 + y^2)$ considered in Example 1.8 above; moreover, it constitutes a challenging problem on distinguishing between center and focus, posed by Poincaré a century ago and still open. A similar though apparently simpler problem was recently studied in [**BFY3**].

The third, even more difficult, problem is related to the fact that a polynomial vector field can be integrable but not Hamiltonian: it is sufficient that the corresponding form Ω possess an integrating factor. A typical example is that of *Darboux integrable* vector fields. Let $\lambda_1, \dots, \lambda_n \in \mathbb{R}$ be a collection of real numbers and $H_1, \dots, H_n \in \mathbb{R}[x, y]$ polynomials. The rational 1-form

$$\Omega' = \sum_{i=1}^{n} \lambda_i \frac{dH_i}{H_i} = \frac{dF}{F}, \qquad F = H_1^{\lambda_1} \cdots H_n^{\lambda_n},$$

determines a conservative (integrable) singular foliation, since its null spaces $\{\Omega' = 0\}$ are tangent to the real level curves $\{F(x, y) = t\} \subset \mathbb{R}^2$ which are all closed unless they begin and end at infinity. The form Ω' is not polynomial (only rational), but the form $\Omega = R\Omega'$ already will be if $R = \gcd(H_1, \dots, H_n) \in \mathbb{R}[x, y]$ is the common divisor of all H_i.

Appearance of limit cycles in polynomial perturbations of the form

$$\Omega + \varepsilon\omega = 0, \quad \omega = p\,dx + q\,dy, \ p, q \in \mathbb{R}[x, y],$$

is determined by the same mechanisms as in the Hamiltonian case, in particular, the first variation of the Poincaré displacement map, measured in units of $\ln F = \sum \lambda_i \ln H_i$, is equal to the integral

$$I(t) = \oint_{F=t} \frac{1}{R(x, y)}\,\omega$$

of the *rational* 1-form $R^{-1}\omega$ along closed level curves of the *transcendental* function F.

As was noted on several occasions by V. Arnold, the problem on Abelian integrals should also be posed for such "pseudo-Abelian" integrals. Nevertheless, there are practically no results of a general nature pertinent to this problem, even in the simplest cases $n = 2$ and $n = 3$. In particular, *all* results described below (derivation of the Picard–Fuchs system, monodromy properties etc.) fail for this generalized class of integrals.

1.4. Finiteness problems for Abelian integrals: tangential Hilbert problem. Despite its limited nature, the connection between limit cycles of polynomial vector fields and isolated zeros of Abelian integrals justifies formulation of several finiteness problems for the latter. As was suggested, we refer to the problem of computing or estimating from above the number of zeros of Abelian integrals as the tangential Hilbert problem, distinguishing between several finiteness types of this problem according to the paradigm laid out before.

1.4.1. *Individual finiteness.* For any fixed combination of the Hamiltonian H and the 1-form ω the Abelian integral appearing in (8) as a function of the real variable t can be shown to be real analytic except for finitely many values of t. According to Section 1.1, it may have only finitely many isolated zeros unless they accumulate to one of these exceptional values. The *individual finiteness problem for Abelian integrals* is to prove that such accumulation is in fact impossible. This assertion turns out to be a simple corollary to a general theorem describing ramification of Abelian integrals after analytic continuation into the complex domain. The key step in the proof of individual finiteness for Abelian integrals is the following representation.

LEMMA 3. *Any Abelian integral near any exceptional point t_* admits a representation of the form*

$$I(t) = \sum_{r,k} h_{kr}(t)\,(t - t_*)^r \ln^{k-1}(t - t_*) \tag{12}$$

with finitely many terms, where all exponents r are rational numbers, the powers k take only finitely many natural number values, and all functions h_{kr} are real analytic at the point $t = t_*$.

This lemma is explained in Section 2.1. As soon as the representation (12) is established, the rest of the proof is as described in Example 1.2. It remains only to notice that the convergence of the expansion (12) is crucial: if all terms in this expansion are zero, then the integral itself is identically zero and hence has no isolated roots at all (compare with the Remark in Section 1.2.3).

1.4.2. *Existential finiteness.* Abelian integrals depend on the choice of the Hamiltonian and the form; in order to make the parameter space finite-dimensional, one has to restrict their degrees. Then the coefficients of the Hamiltonian H and the form ω become the natural parameters of the problem.

The parameter space is intrinsically compact: indeed, replacing H and ω by $c_1 H$ and $c_2\omega$ with c_1, c_2 nonzero constants, clearly does not affect the number of isolated roots of the corresponding integral. Hence the parameter spaces are in fact projective spaces of appropriate dimensions.

Fix any two integers n and d and consider all Hamiltonians of degree $\leqslant n$ and all polynomial forms of degree $\leqslant d$. The existence of *uniform* upper bounds for the number of isolated zeros of the Abelian integrals, subject to the above restrictions on the degrees, was proved by Varchenko [**V**] and Khovanskiĭ [**K1**].

THEOREM 3 (Varchenko – Khovanskiĭ, 1984). *For any n, $d < +\infty$, the number of ovals $\gamma \subset \{H = \text{const}\}$ yielding isolated zeros to the Abelian integral $\oint_\gamma \omega$ is bounded by a constant $N = N(n,d)$ uniformly over all Hamiltonians of degree $\leqslant n$ and all polynomial 1-forms of degree $\leqslant d$.*

One key tool in their proof is again the same Lemma 3 properly generalized to cover the parametric construction. The second principal ingredient is the *Pfaffian elimination* technique [**K2**]. The latter is a method allowing one to reduce the question on zeros of functions involving real powers, logarithms, arctangents and other functions that can be defined using Pfaffian equations with algebraic right-hand sides to the question on zeros of certain auxiliary systems of equations involving only algebraic functions. Application of the generalized Lemma 3 allows one to derive the existential finiteness theorem for Abelian integrals, using in the standard way the compactness and localization arguments, from the similar existential finiteness assertion for convergent multivariate expressions of the form (12). The latter assertion can in turn be derived, using the Pfaffian elimination technique, from the existential finiteness theorem for analytic families of functions as introduced in Section 1.1. This last case was already settled by Theorem 1, and this completes the proof.

However, since both Theorem 1 and its corollary, Theorem 2, do not give any information on the *number* of isolated roots of functions, Theorem 3 inherits this quality as a purely existential statement.

1.4.3. *Constructive finiteness.* The simplest (trivial) case in which constructive bounds for the tangential Hilbert problem can be easily produced is the case of quadratic Hamiltonians. In this case all Abelian integrals can be explicitly computed, being actually *polynomial* functions of t.

The constructive finite cyclicity problem for Abelian integrals is also a relatively simple assertion: the maximal multiplicity of an isolated zero of an Abelian integral admits an upper bound in terms of $\deg\omega$ and $\deg H$ found by P. Mardešić in [**M**].

Historically the first nontrivial problem concerning zeros of Abelian integrals appeared in connection with what later became known as the Takens – Bogdanov bifurcation. In order to prove that no more than one limit cycle appears by deformation of the generic cuspidal singular point on the plane, one has to verify that, for any 1-form $\omega = (\alpha + \beta x)y\,dx$, its integral over the closed ovals of the Hamiltonian $H(x, y) = y^2 + x^3 - 3x$ has no more than one isolated zero. R. Bogdanov in [**B1**]

TABLE 2. Constructive tangential Hilbert 16th problem: partial synopsis

Hamiltonian H	Degree of the form: Low degree ($\deg \omega \leqslant 2$)	Degree of the form: Arbitrary $d = \deg \omega$ (asymptotic)	Degree of the form: Arbitrary degree (constructive)
Quadratic: $H = x^2 + y^2$	Integrals are polynomial functions of t		
Elliptic $H = y^2 + x^3 - 3x$	$\leqslant 1$ root Petrov [**P1**]		nonoscillation theorem Petrov [**P2**]
General cubic H	numerous results in particular, Gavrilov [**G3**]		$5(d + 2)$ Horozov – Iliev [**HI**]
Hyperelliptic $H = y^2 + P_n(x)$	—	Petrov [**P3**]	tower function, Novikov – Yakovenko [**NY3**]
General n**th degree polynomial**	—	$C(n)d + O(1)$, Khovanskiĭ (unpublished)	???

Dashes in the cells mark irrelevant (artificial) problems. Blank cells indicate that the corresponding problem possesses no specific (more accurate) solution other than the one implied by stronger versions appearing to the right or down in the table.

proved this claim, achieving one of the first results on Abelian integrals. Later his proof was considerably simplified by Ilyashenko in the paper [**I3**] which introduced some very important tools, among them the idea of complexification of the Abelian integrals.

There are numerous studies treating other low-degree cases that mostly appeared in connection with bifurcations of polynomial vector fields of low degrees. Starting from the paper by Bogdanov, this direction was pursued, among others, by F. Dumortier, A. Gasull, L. Gavrilov, F. Girard, E. Horozov, I. Iliev, Yu. Ilyashenko, A. Jebrane, B. Li, C. Li, J. Llibre, P. Mardešić, G. Petrov, R. Roussarie, C. Rousseau, Z. Zhang, Y. Zhao, H. Zoladek, to mention only some names and the most recent works. Among these accurate bounds, the following result is remarkable for its succinct formulation and difficult proof.

THEOREM 4 (Gavrilov – Horozov – Iliev [**G3**]). *For any cubic Hamiltonian H with four distinct critical values, and any quadratic 1-form ω, the corresponding Abelian integral has no more than two isolated roots.*

Moreover, in the perturbation (7) *no more than* 2 *limit cycles may appear, including those born from separatrix polygons.*

This settles the localized version of the Hilbert 16th problem for quadratic vector fields arbitrary close to Hamiltonian quadratic vector fields with the specified Hamiltonians (cubic with 4 distinct critical values). The bound is accurate.

Another remarkable result is due to G. Petrov [**P2**] who studied completely the *elliptic case* $H(x, y) = y^2 + P_3(x)$, where $P_3 \in \mathbb{R}[x]$ is a cubic univariate polynomial: by affine transformations one can always reduce P_3 to the form $P_3(x) =$

$x^3 - 3x$. Petrov proved that the Abelian integrals of forms of arbitrary degree form a *nonoscillating* (in other language *disconjugate*, or *Chebyshev*) family: *the number of isolated roots never becomes equal to or exceeds the dimension of the linear space of all such integrals.*

REMARK. One can easily verify that any linear space spanned by d linear independent analytic functions $f_1(t), \dots, f_d(t)$ contains a nontrivial linear combination $c_1 f_1 + \dots + c_d f_d$ exhibiting a root of multiplicity $d-1$ at any preassigned point.

Methods introduced by Petrov were further elaborated and refined. The following bound obtained by E. Horozov and I. Iliev, though not sharp, covers the case of Abelian integrals of arbitrary polynomial forms over level curves of any cubic Hamiltonians.

THEOREM 5 ([**HI**]). *For any cubic Hamiltonian $H(x, y)$ the Abelian integral of a form of degree d cannot have more than $5(d+2)$ isolated zeros.*

By similar methods the *quartic* Hamiltonians (of degree 4) with *elliptic* level curves were studied in [**GJ**]. The case of quartic elliptic Hamiltonians perturbed within the Liénard equation was a subject of four recent preprints of C. Li and F. Dumortier.

1.4.4. *Asymptotic bounds.* Results of yet another type, asymptotic bounds, take advantage of the asymmetry of the roles played by 1-forms and Hamiltonians. Unlike the original Hilbert 16th problem, in which the coefficients of the polynomials P, Q all enjoy equal rights as parameters, the roles of the Hamiltonian H and the form ω in the definition of the Abelian integrals are fairly different (e.g., the integral depends on ω linearly, whereas even small variations of H can result in drastic changes of the domain of definition). Thus it makes complete sense to separate these two parameters and study first the dependence on ω, treating H as fixed ("individual").

The corresponding "semiconstructive" problem was addressed in a number of recent publications. First, very excessively large (double exponential in d) upper bounds were obtained in [**IY1**] and almost immediately improved to simple exponential expressions in [**NY1**] by Ilyashenko, Novikov and the author. The ultimate result in this direction, a *linear* upper bound of the form $C_1(n) \cdot d + C_2(n)$, was obtained by Khovanskiĭ (unpublished) using some ideas developed earlier by G. Petrov [**P3**]. Here the constant $C_1(n)$ is absolutely explicit (e.g., does not exceed a double exponential of n), while the second constant $C_2(n)$ is purely existential though uniform over all Hamiltonians of degree $\leqslant n$.

1.4.5. *Hyperelliptic integrals.* The only particular case covering Hamiltonians of arbitrarily high degree, for which a constructive solution of the tangential Hilbert problem is known is that of *hyperelliptic* Hamiltonians,

$$H(x,y) = y^2 + P_{n+1}(x), \quad P_{n+1} = x^{n+1} + a_1 x^{n-1} + \dots + a_{n-1}x + a_n. \tag{13}$$

Singular points of the corresponding Hamiltonian system correspond to critical points of the polynomial P, called the *potential.* The reason why the class of hyperelliptic polynomials is especially simple one can vaguely attribute to the fact that, from many points of view, hyperelliptic polynomials behave like their univariate potentials. In particular, this concerns the topology of the bundles defined by complexification of H.

Under the addition assumption (believed to be technical, though it occurs independently and persistently in several related problems) that *all critical points of the potential P are real*, the tangential Hilbert problem turns out to be *constructively solvable*. More precisely, as shown by D. Novikov and the author [**NY3**], there exists an algorithm defining an elementary function $C(n,d)$ of two integer arguments n and d such that, for any form of degree $\leqslant d$ and any hyperelliptic Hamiltonian n of degree $\leqslant n$ having only real critical points, the corresponding hyperelliptic Abelian integral has no more than $C(n,d)$ isolated roots.

The algorithm involves several nested inductive constructions, resulting in an *extremely excessively large* bound: it is given by a *tower function* (an iterated exponential) of height greater than 5 but probably smaller than 10.

The proof is based on the fact that Abelian integrals satisfy a system of first order linear ordinary differential equations with rational coefficients, called the *Picard – Fuchs* system.

1.4.6. *Quantitative theory of ordinary differential equations as a tool for the constructive solution of the tangential Hilbert problem.* The method used for constructive solution of the tangential Hilbert problem in the hyperelliptic case is fairly general. In [**Y2**] one can find an introduction to the general theory allowing one to investigate zeros of functions defined by ordinary differential equations with polynomial and rational coefficients. The basics of this theory were developed in a series of joint works by D. Novikov and the author; they are briefly recalled below.

The principal goal of these lecture notes is twofold. First, we show that the Picard – Fuchs system of differential equations can be *explicitly* written down for an arbitrary generic Hamiltonian (not necessarily a hyperelliptic one), at least at the price of a certain *redundancy*. Some additional information may be extracted from the explicit derivation procedure. In particular, we show that for almost any Hamiltonian H (chosen for convenience of degree $n+1$):

(1) there exist μ different Abelian integrals $I_k = \oint \omega_k$ of monomial 1-forms $\omega_1, \dots, \omega_\mu$, $\mu = n^2$, satisfying together a *Fuchsian system* of linear ordinary differential equations with rational coefficients of the form

$$\frac{dx}{dt} = A(t)x, \quad A(t) = \sum_{j=1}^{\mu} \frac{A_j}{t - t_j}, \tag{14}$$

$$x \in \mathbb{C}^\mu, \quad A_j \in \operatorname{Mat}_{\mu\times\mu}(\mathbb{C}), \quad t \in \mathbb{C}, t_1, \dots, t_\mu \in \mathbb{C},$$

such that

(2) the integral $I(t)$ of any other form ω can be represented as a linear combination of the integrals $I_1, \dots, I_\mu$,

$$I(t) = q_1(t)I_1(t) + \cdots + q_\mu(t)I_\mu(t), \tag{15}$$

with polynomial coefficients $q_j \in \mathbb{C}[t]$, $\deg q_j \leqslant \deg \omega / \deg H$.

The residue matrices A_j can be sufficiently completely described, in particular, upper bounds on their norms can be placed.

The second goal is to introduce several results on the number of zeros of functions defined by systems of ordinary linear equations with polynomial and rational coefficients, and polynomial combinations of such functions. We start with the simplest case of one nth order linear equation and show that isolated zeros of its solutions can be described in terms of the magnitude of coefficients of this equation. In the simplest real case this is a classical theorem of de la Vallée Poussin [**dlVP**], which we generalize for the complex analytic context. Afterwards we study systems of polynomial ODE's in real and complex space of arbitrary dimension; here for the first time appear explicit bounds given by tower functions. Finally, we discuss the case of Fuchsian systems of the form (14) and show that, under certain natural restrictions on the monodromy group of this equation, a global upper bound on the number of zeros of solutions can be given in terms of norms of the residue matrices A_j.

One can hope that a combination of these two techniques ultimately would allow one to construct an explicit bound for the tangential Hilbert problem in the general case, filling the right bottom corner of the Table 2.

2. Abelian integrals and differential equations

In this section we explain the connection between Abelian integrals and linear ordinary differential equations.

2.1. Complexification of Abelian integrals: topological approach.

2.1.1. *General scheme.* We recall here the basic construction of complexification of Abelian integrals. All details can be found in various textbooks, [**AGZV**] being the principal source.

A bivariate polynomial $H \in \mathbb{C}[x, y]$ defines a map from $\mathbb{C}^2$ to $\mathbb{C}^1$ with the preimages of points being affine algebraic curves. It turns out that the map H is a *topological bundle* over the complement to finitely many points $\Sigma = \{t_1, \dots, t_r\} \subset \mathbb{C}$. This allows one to identify in a canonical way the homology groups of all fibers $H^{-1}(t)$ sufficiently close to $H^{-1}(t_*)$, which in turn allows one to extend integrals over 1-cycles of polynomial 1-forms as complex analytic functions of t in $U = \mathbb{C} \setminus \Sigma$, ramified over the singular locus Σ. Geometrically this can be expressed as introducing a locally flat connexion on the (co)homology bundles over the punctured sphere, see Section 2.1.3.

From the same topological arguments it follows that for any polynomial 1-form ω the linear space generated by integrals $\oint_{\delta_i} \omega$ over any family $\delta_1(t), \dots, \delta_\mu(t)$ of cycles forming a basis in the first homology group of the fiber $X_t = H^{-1}(t)$ is invariant under analytic continuation along the loops avoiding the locus Σ. The monodromy group consisting of all automorphisms of this space occurring as the result of continuation over all loops is independent of the choice of the form ω and depends only on the Hamiltonian H.

These fairly general topological considerations already imply that the Abelian integrals form a finitely generated module over the ring of polynomial functions of t, and generators of this module satisfy a system of first order linear ordinary differential equations with rational (in t) coefficients, called *Picard – Fuchs* equations.

However, in order to apply methods described in subsequent sections it is necessary to obtain *quantitative* characteristics of these equations, in particular their dimension, degree and the magnitude of coefficients. Part of this information (e.g., the dimension) can be easily achieved from the above construction. To obtain upper bounds on the degrees some other more subtle considerations are required, but it is practically impossible to derive bounds for the coefficients using only topological arguments as above.

The current section contains a brief exposition of well-known facts leading to derivation of Picard – Fuchs equations and representation of the space of Abelian integrals as the Picard – Vessiot extension. In the next section we suggest an alternative approach based on elementary algebraic considerations, which allows one to derive explicitly the Picard – Fuchs system at the price of a certain redundancy.

2.1.2. *Topological bundles defined by proper maps.* Let $f\colon M \to N$ be a smooth map between two manifolds. Recall that a point $b \in N$ is a regular value for f if the rank of the differential $f_*\colon T_xM \to T_bN$ is maximal (equal to $\dim N$) at all points $x \in X_b$ of the preimage. The complement to the set of regular values consists of *critical values* and is denoted $\operatorname{crit} f$.

LEMMA 4. *If b is a regular value of a proper map $f\colon M \to N$, then there exists a neighborhood $U \ni b$ such that for all $y \in U$ the preimages $X_y = f^{-1}(y) \subset M$ are diffeomorphic to $X_b = f^{-1}(b)$ and hence $f^{-1}(U) \simeq X_b \times U$.*

PROOF. Consider an arbitrary vector $v_0 \in T_bN$ and embed it into a vector field v on N. We claim that in a sufficiently small neighborhood of the preimage X_b one can construct a smooth vector field w such that $f_*w = v$, that is, w and v are f-related. Such a field obviously exists near each point $x \in X_b$, since f_* is surjective (and takes the form of a parallel projection in suitably chosen local coordinates by virtue of the theorem on rank). Now it remains to choose a finite covering of X_b by these neighborhoods and patch together the corresponding vector fields, using the appropriate partition of unity.

To conclude the proof, notice that the (local) flows of v and w are conjugate by f (by construction), hence the local flow of w, defined in some neighborhood of X_b, takes the latter preimage into the preimage of the corresponding point y on the flow curve of b. Since the initial vector v_0 can be chosen pointing in any direction, this proves that all sufficiently close preimages are diffeomorphic to each other. □

REMARK. The diffeomorphism between close preimages is not canonically defined, but any two such diffeomorphisms are homotopically equivalent, since the vector field w from the proof of the lemma is homotopically unique.

COROLLARY 1. *If M is compact, then f is a topological bundle over $N \setminus \operatorname{crit} f$.* □

REMARK. The assertion of this lemma for the case $N = \mathbb{R}$ is the centerpiece of the Morse theory.

2.1.3. *Homology bundle and flat connexion on it.* A locally trivial topological bundle $f\colon M \to N$ defines in a canonical way the associated homology bundle over N with the fibers being the homology groups $H_i(X_y, \mathbb{Z})$ (we will be only interested in the case $i = 1$) together with a flat connexion on this bundle. This means that any 1-cycle on any particular fiber X_a can be transported to any other fiber X_b along any path γ connecting a and b in N. Flatness means that continuation along any sufficiently small loop beginning and ending at a returns any 1-cycle to its initial position. However, transport along "long" loops (not contractible in N) can result in a nontrivial transformation of the homology, called the *monodromy transformation.*

Fix a point $a \in N$ and choose a basis $\delta_1, \dots, \delta_\mu$ of 1-cycles in the group $H_1(X_a, \mathbb{Z})$, where μ is the rank of the homology group. Then the monodromy transformation corresponding to any loop γ from the fundamental group $\pi_1(N, a)$ can be described by the corresponding *monodromy matrix* $M_\gamma \in \operatorname{Mat}_{\mu\times\mu}(\mathbb{Z})$ with integral entries m_{ij}:

(16) $$\Delta_\gamma \delta_j = \sum_{i=1}^{\mu} \delta_i m_{ij}, \quad i = 1, \dots, \mu.$$

In the matrix form the monodromy transformation acts on the row vector $\boldsymbol{\delta} = (\delta_1, \dots, \delta_\mu)$ as multiplication by the matrix M_γ from the right, $\Delta_\gamma \boldsymbol{\delta} = \boldsymbol{\delta} \cdot M_\gamma$.

The integer-valued matrix M_γ is obviously invertible. Moreover, the correspondence $\gamma \mapsto M_\gamma$ is a representation of the fundamental group $\pi_1(N, a)$ in $\mathrm{GL}(n, \mathbb{Z})$: for any two loops γ, γ' and their product $\gamma \circ \gamma'$ we have $M_{\gamma\circ\gamma'} = M_{\gamma'} M_\gamma$. This implies, in particular, that $\det M_\gamma = \pm 1$ for all loops γ.

The flat connexion on the homology bundle defines a connexion on the cohomology bundle which makes it possible to compute the covariant derivative of 1-forms: any smooth 1-form ω on the manifold M can be restricted to any fiber X_a and integrated along a continuous (horizontal, locally constant) family of 1-cycles $\delta(y)$, resulting in a function $I(y)$. The result of continuation of this function along closed paths in N is completely determined by the monodromy group and does not depend on the choice of the form ω. More precisely, integrating the form ω over each of the 1-cycles $\boldsymbol{\delta}(y) = \big(\delta_1(y), \dots, \delta_\mu(y)\big)$, we obtain a tuple of continuous functions $I_j(y) = \oint_{\delta_j(y)} \omega$, $j = 1, \dots, \mu$, which after continuation along a path $\gamma \in \pi_1(N, a)$ undergo the transformation

(17) $$\Delta_\gamma \boldsymbol{I} = \boldsymbol{I} \cdot M_\gamma, \quad \boldsymbol{I} = (I_1, \dots, I_\mu), I_j = I_j(y)$$

with the same matrices M_γ independently of the form ω. This basic fact lies at the core of the topological theory outlined below.

2.1.4. *Topological bundles defined by polynomial maps.* Our goal is to apply the previous construction to the polynomial map $H\colon \mathbb{C}^2 \to \mathbb{C}$ considered as a smooth map between smooth manifolds. Since compactness of the preimages is crucial for these arguments, we need to compactify the domain (and the range) of H.

In contrast to the one-dimensional case, it is in general impossible to extend H as a map between the natural compactifications $\mathbb{C}P^2$ and $\mathbb{C}P^1$ respectively, since on the line at infinity $\mathbb{C}P^1_\infty \subset \mathbb{C}P^2$ one has several *points of indeterminacy*: they occur at the intersections between compactified preimages $H^{-1}(t)$ and the line at infinity. Algebraically this can be seen as the indeterminacy of the rational expression $H(1/z, y/z) = P(y,z)/z^d$, $d = \deg H$, at the points where $P(y, 0) = 0$ (at all other points of the infinite line $\{z = 0\}$ one can assign the value $H = \infty$ to this ratio).

The problem can be resolved by blowing up these indeterminacy points, in the same way as blowing up the origin allows one to assign values from $\mathbb{C}P^1 = \mathbb{C} \cup \{\infty\}$ to the rational expression $R(x, y) = y/x$ that is initially indeterminate. After a series of blow-ups at indeterminacy points one arrives at a *compact* two-dimensional complex manifold M and a map (still denoted by H) from M to $\mathbb{C}P^1$, called the *determination* of the initial polynomial map.

Now one can apply Lemma 4. Since the determination $H\colon M \to \mathbb{C}P^1$ is a rational map between compact algebraic varieties, it has only a finite number of critical values $\Sigma_H = \{t_1, \dots, t_s\} \subset \mathbb{C}P^1$ and we conclude that it defines a topological bundle over the complement of these *exceptional values*. Any 1-cycle $\delta(a) \subset H^{-1}(a)$ on a fiber of this bundle can be embedded into a continuous horizontal family $\delta(t)$ of 1-cycles, ramified over the exceptional locus. A polynomial 1-form ω on $\mathbb{C}^2$

extends as a meromorphic 1-form on $\mathbb{C}P^2$ with the polar divisor $\mathbb{C}P^1_\infty$ that after the blowing up becomes an algebraic hypersurface D in M. Let Σ be the union of Σ_H and the critical values of the projection H restricted to D (including the images of the nonsmooth points of D).

The pullback of ω on M can be integrated along the family $\delta(t)$: by the Cauchy–Stokes theorem, the result depends only on the homology class of the cycle. It can be easily shown that the result of this integration is an analytic function of t, ramified over Σ. Its monodromy (the result of analytic continuation along closed loops avoiding the exceptional locus Σ) is as before determined only by H.

This construction proves the following result.

PROPOSITION 1 (cf. with [**AGZV**]). *For any polynomial $H\colon \mathbb{C}^2 \to \mathbb{C}$, any continuous family of 1-cycles $\delta(t)$ on nonsingular level curves $X_t = \{H = t\} \subset \mathbb{C}^2$ and any polynomial 1-form ω on $\mathbb{C}^2$ the Abelian integral $I(t) = \oint_{\delta(t)} \omega$ is an analytic multivalued function ramified over the finite set Σ_H depending only on H.* □

Behavior of the integrals near the ramification locus is relatively tame: it can be seen that any integral can grow no faster than polynomially in $|t - t_j|^{-1}$ as t tends to some $t_j \in \Sigma$ remaining in any sector with the vertex at t_j.

2.1.5. *Picard–Fuchs system.* Let as before $\boldsymbol{\delta}(t) = \big(\delta_1(t), \dots, \delta_\mu(t)\big)$ be a continuous family of 1-cycles forming a basis (frame) of the first homology group of the respective fibers $H^{-1}(t)$, arranged as a row vector. One can show that there exist μ polynomial 1-forms $\omega_1, \dots, \omega_\mu$ such that the *period matrix*

$$X(t) = \begin{pmatrix} \oint_{\delta_1} \omega_1 & \cdots & \oint_{\delta_\mu} \omega_1 \\ \vdots & \ddots & \vdots \\ \oint_{\delta_1} \omega_\mu & \cdots & \oint_{\delta_\mu} \omega_\mu \end{pmatrix}, \qquad \delta_j = \delta_j(t), \tag{18}$$

is not identically degenerate, $\det X(t) \not\equiv 0$. From (17) it follows that

$$\Delta_\gamma X(t) = X(t) M_\gamma, \quad \forall \gamma \in \pi_1(\mathbb{C}P^1 \setminus \Sigma, a). \tag{19}$$

Differentiating the identity (19), we see that the derivative $\dot X(t)$ has the same monodromy (i.e., $\dot X$ is multiplied by the same matrix factors M_γ). Therefore the "logarithmic derivative" $A(t) = \dot X(t) \cdot X^{-1}(t)$ is a single-valued (invariant under all monodromy transformations) meromorphic matrix function having poles of finite order at the points of Σ and eventually at the points of degeneracy of $X(\cdot)$:

$$\Delta_\gamma A(t) = \dot X(t) M_\gamma \cdot M_\gamma^{-1} X^{-1}(t) = \dot X(t) \cdot X^{-1}(t) = A(t)$$

for any loop $\gamma \in \pi_1(\mathbb{C} \setminus \Sigma, t_*)$. From this we conclude that $A(t)$ is a rational matrix function while the period matrix $X(t)$ is a fundamental matrix solution to the system of linear ordinary differential equations with rational coefficients on $\mathbb{C}P^1$,

$$\dot X = A(t) X, \quad A(\cdot) \in \operatorname{Mat}_{\mu\times\mu}\big(\mathbb{C}(t)\big). \tag{20}$$

The common name for various such systems satisfied by Abelian integrals is the *Picard–Fuchs system* (or Picard–Fuchs equation).

Integrals of any other form ω can be expressed as linear combinations of integrals of the framing forms ω_i, $i = 1, \dots, \mu$, with coefficients from the field $\mathbb{C}(t)$ of rational functions. Indeed, multiplying the row vector $\boldsymbol{I}(t) = \big(I_1(t), \dots, I_\mu(t)\big)$,

$I_j = \oint_{\delta_j} \omega$, by $X^{-1}(t)$, we obtain (for the same reasons as above) a single-valued hence rational row-vector function $\boldsymbol{r}(t) = \big(r_1(t), \dots, r_\mu(t)\big)$, that is,

$$\oint_{\delta(t)} \omega = \sum_{j=1}^{\mu} r_j(t) \oint_{\delta(t)} \omega_j, \quad r_j \in \mathbb{C}(t). \tag{21}$$

for any cycle $\delta(t)$ continuously depending on t. Note that the space of functions representable as (21) is closed under derivation.

We summarize this as follows. Recall that a *Picard–Vessiot extension* is a differential field of analytic multivalued functions obtained by adjoining to the field $\mathbb{C}(t)$ all components of a fundamental matrix solution of a system of linear ordinary differential equations with rational coefficients.

THEOREM 6. *Abelian integrals belong to a Picard–Vessiot extension for some system of linear ordinary differential equations with rational coefficients.*

Later we discuss this formulation and relevant issues in more detail. However, it is important to stress here that neither the entries of the rational matrix $A(t)$ nor the rational coefficients $r_j(t)$ can be computed explicitly from the above construction without additional considerations.

2.2. Hamiltonians transversal to infinity.

2.2.1. *Definition.* For an arbitrary Hamiltonian H, even location of the ramification points t_j is difficult to describe without effectively resolving all the indeterminacy points at infinity. However, under some natural and generic assumptions one may guarantee that no new critical points will appear after compactification and blowing up the indeterminacy points on the line at infinity $\mathbb{C}P^1_\infty \subset \mathbb{C}P^2$. This will immediately imply that the set Σ of exceptional values must be a subset of $\operatorname{crit} H$, the set of critical values corresponding to *finite* critical points in $\mathbb{C}^2$ only.

DEFINITION 5. A polynomial $H \in \mathbb{C}[x, y]$ is said to be *transversal to infinity* if its principal homogeneous part $L = \sum_{i+j=d} a_{ij} x^i y^j$, $d = \deg H$, factors as a product of pairwise different linear forms.

Equivalent conditions follow.

(1) The principal homogeneous part has an isolated critical point at the origin;
(2) The partial derivatives $\partial L/\partial x$, $\partial L/\partial y$ are mutually prime;
(3) H has exactly $\mu = (\deg H - 1)^2$ critical points in $\mathbb{C}^2$ if counted with multiplicities;
(4) Each level curve $\{H = t\}$ intersects transversely the line at infinity $\mathbb{C}P^1_\infty \subset \mathbb{C}P^2$ after projective compactification.

2.2.2. *Topology of polynomials transversal to infinity.* We prove now that, for H transversal to infinity, $\Sigma = \operatorname{crit} H$.

PROPOSITION 2. *A polynomial* $H\colon \mathbb{C}^2 \to \mathbb{C}^1$ *transversal to infinity is a topological bundle over the set* $\operatorname{crit} H$ *of critical values of* H.

Instead of proving this by resolving the indeterminacy points at infinity, one may modify the proof of Lemma 4 sketched above, and construct the vector field w near an arbitrary point at infinity $p \in \mathbb{C}P^1_\infty$ on $\mathbb{C}P^2$ with the property $H_* w = \partial/\partial t$.

PROOF. It will be shown below in Section 2.3.1 that if H of degree $n+1$ is transversal to infinity, then there exist two polynomials $a, b \in \mathbb{C}[x,y]$ of degree $n-1$ such that

$$a\frac{\partial H}{\partial x} + b\frac{\partial H}{\partial y} = x^{2n-1} + \cdots,$$

where the dots stand for a polynomial of degree $\leqslant 2n-2$ in x, y. One can easily check that the rational vector field

$$w = \frac{a(x,y)}{x^{2n-1}+\cdots}\frac{\partial}{\partial x} + \frac{b(x,y)}{x^{2n-1}+\cdots}\frac{\partial}{\partial y},$$

H-related to the field $\partial/\partial t$, in the chart $(1/x, y/x)$ is regular (smooth) *on the line at infinity* $1/x = 0$ (more precisely, on its affine part covered by this chart). The other affine part is covered by the field that is obtained in a similar way from solution of the equation $a\partial H/\partial x + b\partial H/\partial y = y^{2n-1} + \cdots$.

Thus near each point of the compactified level curve $\overline{X}_a \subset \mathbb{C}P^2$ one has a smooth vector field H-related to $\partial/\partial t$ in the finite part $\mathbb{C}^2 \subset \mathbb{C}P^2$ (in particular, this implies that this field vanishes at all points of indeterminacy of H on $\mathbb{C}P^1_\infty$). The rest of the proof is the same as in Lemma 4. □

2.2.3. *Module of Abelian integrals, Gavrilov and Novikov theorems.* For Hamiltonians transversal to infinity, the constructions of Section 2.1 can be further refined. In particular, the choice of the forms $\omega_1, \ldots, \omega_\mu$ can be made much more explicit.

LEMMA 5 (L. Gavrilov [**G2**]). *Let H be a Hamiltonian of degree $n+1$ transversal to infinity, with distinct critical values $t_1, \ldots, t_\mu$, $\mu = n^2$. Then one can choose n^2 monomial 1-forms $\omega_1, \ldots, \omega_\mu$ of degrees $\leqslant 2n$ so that the respective period matrix has the determinant* $\det X(t) = c(t-t_1)\cdots(t-t_\mu)$ *with $c \neq 0$.*

In this assertion we use the convention on degrees of polynomial k-forms formulated in Section 2.3.1. As a corollary, one can derive the following result that refines the assertion of Theorem 6.

COROLLARY 2. *The Abelian integral of a 1-form ω of degree d can be represented as*

$$\oint_{\delta(t)} \omega = \sum_{j=1}^{\mu} p_j(t) \oint_{\delta(t)} \omega_j, \tag{22}$$

$$p_j(t) \in \mathbb{C}[t], \quad \deg\omega_j + \deg H \cdot \deg p_j \leqslant \deg\omega.$$

In other words, Abelian integrals constitute a module over the ring $\mathbb{C}[t]$ that is generated by integrals of the basic forms ω_j.

The constant c from Lemma 5 depends on the choice of the monomial forms and the Hamiltonian. Its value was explicitly computed by A. Glutsuk [**G**] following some ideas of Yu. Ilyashenko, and a simple elementary proof of the inequality $c \neq 0$ for an appropriate choice of the monomial forms was obtained by D. Novikov [**N2**]. In the same paper [**N2**] it is proved, using some of the methods described below, that the period matrix satisfies a system of linear ordinary differential equations

$$\dot{X} = \frac{1}{(t-t_1)\cdots(t-t_\mu)}P(t)X, \quad P(t) = \sum_{j=0}^{\mu} t^j P_j, \tag{23}$$

with a matrix polynomial $P(t)$ of degree μ, in general having Fuchsian singularities at all points t_j of the ramification locus Σ, but a *non*-Fuchsian singularity at $t = \infty$.

As yet another corollary, one can derive a Picard – Fuchs system for the period matrix. Let ω_j be as in Corollary 2. Consider the closed 2-forms $dH \wedge \omega_j$ and let Ω_j be any collection of polynomial primitives satisfying the conditions

$$d\Omega_j = dH \wedge \omega_j, \quad j = 1, \dots, \mu.$$

Each Ω_i can be expanded as in Corollary 2, yielding an identity between the integrals,

$$\oint \Omega_i = \sum_{j=1}^{\mu} p_{ij}(t) \oint \omega_j$$

valid for any choice of a continuous family of cycles of integration $\delta(t)$. On the other hand, as will be shown in Section 2.3.2, the derivative of each integral $\oint \Omega_i$ is exactly the integral $\oint \omega_i$. Differentiating the above identities, we arrive at the matrix differential equation

$$X = \dot{P}X + P\dot{X}, \quad P = P(t) = \|p_{ij}(t)\| \in \operatorname{Mat}_{\mu\times\mu}(\mathbb{C}[t])$$

with a matrix polynomial $P(t)$ of some known degree. If required, this identity can be resolved to the form (20).

2.2.4. *Commentaries.* The proof of Lemma 5 and Corollary 2 is based on a finer than before analysis of topology of the bundle H for polynomials transversal to infinity. In particular, in assumptions of the lemma, one can choose a special framing of the homology bundle by *vanishing cycles* $\delta_j(t)$, represented by loops on the preimage X_t that shrinks to a point when $t \to t_j$ (a special precaution is required to avoid problems with multivaluedness). For such choice of the cycles, the period matrix $X(t)$ must have a vanishing column at each of the points t_j, for any collection of the framing forms. Next, in this case the determinant $\det X$ is a single-valued function that therefore must be a polynomial (being locally bounded everywhere on $\mathbb{C}$). Its growth as $t \to \infty$ depends on the degrees of the framing forms, since the "size" of the cycles $\delta_j(t)$ grows in a known way (depending only on H).

All this implies that the determinant of any period matrix $\det X(t)$ is a polynomial in t divisible by $\prod_{j=1}^{\mu}(t - t_j)$. The coefficients $p_j(t)$ of representation (22) can be found by solving a system of linear algebraic equations. When solved by the Cramer rule, this system yields $p_j(t)$ as a ratio of two determinants, the determinant of the period matrix for the basic forms $\omega_1, \dots, \omega_\mu$ in the denominator, and that of a similar period matrix for the collection of 1-forms with ω_j replaced by ω in the numerator. By Lemma 5, these ratios are polynomials in t and their degrees can be easily majorized in terms of $d = \deg \omega$.

2.2.5. *Multiplicity of the roots of Abelian integrals: constructive finite cyclicity.* A similar construction allows one to majorize the order of a zero of any Abelian integral, at least under the assumption that H is transversal to infinity and has only Morse critical points [**M**]. Instead of the period matrix X, consider the matrix function $J(t)$ whose entry $J_{ij}(t)$ is the $(i-1)$-st derivative of the integral $I_j(t) = \oint_{\delta_j(t)} \omega$.

For the same reasons as before, its determinant $w(t) = \det J(t)$, the Wronskian of the integrals $I_1, \dots, I_\mu$, is a single-valued hence rational function of t. Its poles may occur only at the points t_j and $t = \infty$. The assumption on finite singular points implies that all integrals $I_j(t)$ have at worst logarithmic growth near each t_j, and this growth rate allows for differentiation so that $I_j^{(i-1)}(t)$ grows no faster than

$|t-t_j|^{1-j}$ as $t \to t_j$ without spiraling. These estimates imply an upper bound on the total order of all poles of w at all finite points. The growth rate of w as $t \to \infty$ depends on $\deg \omega$ and can be easily estimated. This gives an upper bound on the degree $\nu = \deg w(t)$ of the rational function (the total number of its poles on $\mathbb{C}P^1$ including those at infinity). This degree is obviously an upper bound for the order of any nontrivial zero of w. Since the order of the Wronskian w is equal to μ, the number $\nu + \mu - 1$ is an upper bound for the order of any root of any integral $I_j(t)$ at any point $t \neq t_j$.

2.2.6. *Reservations.* Despite more detailed constructions and more accurate considerations, the approach based only on topological ideas cannot provide many important data. For example, even the "constant" c from Lemma 5 depends in a rather nontrivial way on both H and the choice of the framing forms ω_j, see [**G**] and [**N2**]. Among other things, this means the practical impossibility of majorizing the polynomial coefficients p_j in (22). The same refers to the derivation of the Picard–Fuchs system: after an accurate computation, it can be reduced to the form determined by three constant matrices P_1, P_2, P_3,

$$(P_0 + tP_1)\dot{X} = P_2 X, \quad P_i \in \operatorname{Mat}_{\mu\times\mu}(\mathbb{C})$$

but no bounds on the norms of these matrices (or their inverses) can be obtained except for some especially simple cases.

In short, it is combinatorial parameters like degrees, dimensions, ranks and so on that can be more or less easily derived from even the most explicit topological constructions. In contrast, all magnitude-like parameters (norms, absolute values, diameters of point sets, etc.), require additional arguments. Some of them can be obtained using rather elementary algebraic considerations.

2.3. Elementary derivation of the Picard–Fuchs system. In this section we derive yet another Picard–Fuchs system by explicit linear algebraic considerations. The advantage of this approach (besides its transparency) is that it allows one to bound explicitly the magnitude of coefficients of the system. In addition, the Picard–Fuchs system obtained this way possesses a nice *hypergeometric* form, exhibiting only Fuchsian singularities (though this fact has not yet been fully exploited). However, this transparency and explicitness is achieved at the price of a certain redundancy: the dimension of the system obtained this way is twice as big as the minimal possible one.

2.3.1. *Division by the gradient ideal.* Division with remainder by an ideal can be expressed in the language of polynomial differential forms. In what follows we consider k-forms Λ^k, $k = 0, 1, 2$, with polynomial coefficients on the plane $\mathbb{C}^2$. The degree of a k-form *by definition* is the maximum of degrees of its coefficients, plus k. Under such a convention, $\deg(\xi \wedge \eta) \leqslant \deg \xi + \deg \eta$ for all admissible combinations of ranks of ξ and η between 0 and 2, and also

$$\deg d\omega \leqslant \deg \omega \quad \forall \omega \in \Lambda^k,$$

for any rank $k = 0, 1$. The linear space of forms of rank k and degree d will be denoted by Λ^k_d, and we denote $\Lambda^k_{\leqslant d} = \Lambda^k_0 + \Lambda^k_1 + \cdots + \Lambda^k_d$.

Consider $\omega = a\,dx + b\,dy \in \Lambda^1$ with only isolated singularities. This implies that the ideal (a, b) generated by the polynomials $a, b \in \mathbb{C}[x, y] \simeq \Lambda^0$ has finite codimension μ in $\mathbb{C}[x, y]$, that is, there exist μ polynomials $r_1, \ldots, r_\mu \in \mathbb{C}[x, y]$ such that any polynomial q from this ring can be represented as

$$q = av - bu + \sum_1^{\mu} c_i r_i, \qquad u, v \in \mathbb{C}[x,y],\ c_1, \dots, c_\mu \in \mathbb{C}.$$

Introducing 2-forms $\Omega = q\,dx \wedge dy$, $R_i = r_i\,dx \wedge dy$, $i = 1, \dots, \mu$, and the 1-form $\eta = u\,dx + v\,dy$ the above identity can be rewritten as *division with remainder*,

$$\forall \Omega \in \Lambda^2 \quad \exists \eta \in \Lambda^1 : \quad \Omega = \omega \wedge \eta + R, \quad R = \sum_{i=1}^{\mu} c_i R_i \in \Lambda^2. \tag{24}$$

In particular, any 2-form Ω can be divided with remainder by the differential dH of any polynomial $H \in \Lambda^0$. Note that the space of remainders may be arbitrarily enlarged if necessary: the uniqueness of the division (24) will then be lost, but in exchange one may get better norms of the ratio and remainder.

In general the procedure of division can be very delicate. However, if H is transversal to infinity, then one can easily describe the outcome, explicitly majorizing the *degrees* of the remainder $R \in \Lambda^2$ and the incomplete ratio $\eta \in \Lambda^1$.

LEMMA 6 (see **[NY5]**). *If the polynomial $H \in \Lambda^0$ of degree $n+1$ is transversal to infinity, then any 2-form Ω can be divided by dH with the incomplete ratio η of degree $\leqslant \deg \Omega - \deg H$ and the remainder R of degree $\leqslant 2n$.*

PROOF. If $L \in \Lambda^0$ is a homogeneous polynomial of degree $n+1$ without multiple linear factors, then the map between subspaces of homogeneous forms,

$$\mathfrak{J} = \mathfrak{J}_L \colon \Lambda^1_n \to \Lambda^2_{2n+1}, \qquad \eta \mapsto dL \wedge \eta, \tag{25}$$

is an isomorphism. Indeed, in the bases consisting of all monomial forms of the given degrees, the matrix of $\mathfrak{J}$ is the *Sylvester matrix* whose determinant is the *resultant* of the two partial derivatives L_x and L_y. The assumption on H implies that this resultant is nonzero.

Therefore any homogeneous form Ω of degree exactly $2n+1$ is divisible by dL. Any monomial form of degree greater than $2n+1$ can be represented as a monomial 1-form of degree $2n+1$ times a monomial function and hence is divisible by dL *without remainder* with the same relation between the degrees,

$$\deg \Omega \geqslant 2n+1 \implies \Omega = dL \wedge \eta, \quad \deg \eta \leqslant \deg \Omega - \deg L.$$

Applying this observation to all homogeneous components of a 2-form $\Omega = \Omega_0 + \cdots + \Omega_{2n} + \Omega_{2n+1} + \cdots + \Omega_d = R + \Omega_{2n+1} + \cdots + \Omega_d$, we prove the assertion of the lemma for homogeneous polynomials.

To divide a 2-form Ω of degree $\geqslant 2n+1$ by a nonhomogeneous differential $dH = dL + \xi$, where L is the principal homogeneous part of H, $\deg \xi < \deg L$, we divide it by dL first, and then transform the result as follows,

$$\Omega = dL \wedge \eta + R = (dH - \xi) \wedge \eta + R = dH \wedge \eta + \Omega',$$
$$\Omega' = R - \xi \wedge \eta, \quad \deg \Omega' \leqslant \max(\deg \Omega - \deg \xi + \deg L, \deg R) < \deg \Omega,$$

reducing division of Ω by dH to division of another form Ω' of strictly inferior degree. Iterating this step, we prove the lemma in the general case. Notice that this is essentially the algorithm of division with remainder of univariate polynomials. □

REMARK. Lemma 6 is an example of *redundant* division. For a Hamiltonian H of degree $n+1$ transversal to infinity, the gradient ideal (ideal of 2-forms divisible by dH) has codimension n^2. Indeed, the codimension is equal to the number μ of critical points of H in $\mathbb{C}^2$, counted with their multiplicities. This latter number is exactly n^2 by virtue of Bézout's theorem, since no critical points are allowed to "escape to infinity" by the assumption on H.

On the other hand, the linear space of bivariate monomials of degree $\leqslant 2n-2$ (the space of 2-forms $\Lambda^2_{\leqslant 2n}$) is $\nu = 2n(2n-1)/2 \approx 2n^2$, roughly twice as big as μ.

The irredundant analog of this theorem can be easily restored if necessary. However, the choice of monomial 2-forms generating the remainder will depend on the principal homogeneous part L of H.

2.3.2. *Gelfand – Leray residue and derivative.*

LEMMA 7. *Let $\omega, \eta \in \Lambda^1$ be two polynomial 1-forms such that*

$$d\omega = dH \wedge \eta.$$

Then for any continuous family $\delta(t)$ of 1-cycles on the level curves $H^{-1}(t)$,

$$\frac{d}{dt}\oint_{\delta(t)} \omega = \oint_{\delta(t)} \eta.$$

The proof in the real case (assuming only smoothness of the forms) can be achieved by integration of $d\omega$ over the annulus between $\delta(t)$ and $\delta(t+\Delta t)$ and passing to the limit as $\Delta t \to 0$.

This formula allows one to differentiate explicitly Abelian integrals of a form ω, expressing the result as an Abelian integral once again if $d\omega$ is divisible by dH. In fact, η can be a *rational* 1-form having all zero residues after restriction to each level curve $H^{-1}(t)$.

EXAMPLE 2.1. Let $H(x,y) = x^2+y^2$ and $\omega = y\,dx$. Then $\oint_{H=t}\omega = -\pi t$ (consider only the real values of t and use the Stokes formula for the circle positively oriented). Clearly, the form $\eta = \frac{1}{2}y^{-1}\,dx$ satisfies the assumption of the lemma, and indeed

$$\oint_{H=t} \eta = \int_0^{2\pi} \frac{1}{2}\cdot\frac{d\cos s}{\sin s} = -\pi.$$

This example helps one to memorize the order of the wedge multiplication in the Gelfand – Leray formula. ◀

2.3.3. *Derivation of the redundant Picard – Fuchs system.* The linear space $\Lambda^2_{\leqslant 2n}$ of possible remainders occurring in the division (24), is spanned by monomial forms $x^r y^s\,dx\wedge dy$, $r+s \leqslant 2n-2$. Denote its dimension $(2n-1)n$ by ν and choose any monomial primitives ω_i, $i=1,\dots,\nu$, so that $d\omega_i$ span the quotient space $\Lambda^2_{\leqslant 2n}/d\Lambda^1_{\leqslant 2n}$ (modulo exact forms). Below we refer to ω_i as the *basic* forms.

Consider the 2-forms $H\,d\omega_i \in \Lambda^2_{\leqslant 3n+1}$ and divide them with remainder by dH:

$$H\,d\omega_i = dH \wedge \eta_i + R_i, \quad i=1,\dots,\nu. \tag{26}$$

By the assertion on the degrees, $\deg\eta_i \leqslant \deg\omega_i + \deg H - \deg dH \leqslant \deg\omega_i \leqslant 2n$, therefore each of the forms can be represented (modulo an exact polynomial form) as a linear combination of the basic forms,

$$\eta_i = \sum_{j=1}^{\nu} b_{ij}\omega_j + dF_i, \quad b_{ij}\in\mathbb{C},\ F_i \in \Lambda^0_{\leqslant 2n}.$$

Similarly, being all of degree $\leqslant 2n$, the remainders $R_i \in \Lambda^2_{\leqslant 2n}$ can be represented as linear combinations of the forms $d\omega_i$:

$$R_i = \sum_{j=1}^{\nu} a_{ij}\, d\omega_j, \qquad a_{ij} \in \mathbb{C}.$$

Let $\delta(t)$ be any continuous family of cycles. Then for any $t \notin \Sigma$ the forms

$$H\, d\omega_i - \sum_{j=1}^{\nu} a_{ij} d\omega_j, \qquad \forall i = 1, \dots, \nu,$$

are all divisible by dH with the ratios being cohomologous to $\sum_{j=1}^{\nu} b_{ij}\omega_j$. Denote

$$X_i(t) = \oint_{\delta(t)} \omega_i, \qquad i = 1, \dots, \nu.$$

Note that integration of a form $H\omega$ over any cycle $\delta \subset \{H = t\}$ yields $t \oint_\delta \omega$, since H is constant on the cycle. Integrating both sides over the oval $\delta(t)$ and using the Gelfand–Leray formula, we arrive at the identities

$$t\dot{X}_i(t) - \sum_{j=1}^{\nu} a_{ij}\dot{X}_j(t) = \sum_{j=1}^{\nu} b_{ij} X_j(t), \tag{27}$$

which means that the *column* vector $(X_1, \dots, X_\nu)$ of Abelian integrals satisfies the system of linear ordinary differential equations

$$(tE - A)\dot{X} = BX, \quad X \in \mathbb{C}^\nu, \quad A, B \in \operatorname{Mat}_{\nu\times\nu}(\mathbb{C}), \tag{28}$$

with the constant matrices $A = \|a_{ij}\|$, $B = \|b_{ij}\|$ as parameters (E is the identity matrix). Writing

$$(tE - A)^{-1} = \frac{1}{\chi(t)} \cdot P(t), \quad P = \sum_{k=0}^{\nu-1} P_k t^k, \ \chi(t) = \det(tE - A), \tag{29}$$

where $P(t)$ is a matrix polynomial of degree $\nu - 1$, the adjugate matrix for $tE - A$, one sees immediately that (28) is a system of linear ordinary differential equations with rational coefficients.

2.3.4. *Hypergeometric systems.* The form (28) is rather specific: for instance, all singular points of this system are Fuchsian (simple poles), including the point at $t = \infty$, provided that the spectrum of A is simple. It would be appropriate to remark here that the residues A_j of the Fuchsian system (13) are invariant under any conformal change of the independent variable t. In the case of hypergeometric systems (28) the point $t = \infty$ is distinguished: the residues A_j at all finite points $t_j \in \operatorname{Spec} A$ have rank 1 for a generic matrix A, whereas the rank of the residue $A_\infty = -\sum_1^\mu A_j$ at infinity is generically full. Thus the natural symmetry group of hypergeometric systems is not the full group of conformal automorphisms of $\mathbb{C}P^1$, but rather the affine group of transformations $t \mapsto at + b$, $a, b \in \mathbb{C}$ fixing the point $t = \infty$. Making an affine transformation transforms the system (28) into the system $(tE + A')\dot{X} = B'X$ with the same matrix $B' = B$ and $A' = a^{-1}(A - bE)$.

In the subsequent sections it will be shown that in order to estimate the number of isolated zeros of solutions to Fuchsian systems, it is sufficient to know the norms of the residue matrices. As follows from the explicit inversion formula (29), norms of the residues can be bounded if the norm $\|A\|$ is bounded from above and pairwise distances between the critical points t_j are bounded *from below*. Of course, choosing a suitable affine transformation as above, one can change the norm $\|A\|$, but at the same rate the distances between the singular points (eigenvalues of A) will be affected. In other words, the norm of the matrix $\|A\|$ should be majorized relative to the spread of its eigenvalues.

2.3.5. *Explicitness and bounds.* The above derivation does not involve any existential assertion: all constructions are completely transparent and allow for explicit bounds, say, on the norms of the matrices A, B from (28). To do this, we introduce the norms on the ring of polynomials, letting $\|p\|$ being the sum of absolute values of all its coefficients. This norm is multiplicative, $\|pq\| \leqslant \|p\| \, \|q\|$, and extends to polynomial k-forms, remaining multiplicative with respect to the wedge product. The exterior derivative is a bounded operator on forms of bounded degrees.

The entries of the matrices A, B appear as coefficients of the linear expansion of a known 1-form in the chosen basis. But since the basic forms ω_i are *monomial* with coefficients equal to 1, to majorize these entries it is sufficient to majorize the norms of the 2-forms R_i (the remainders) and 1-forms η_i (incomplete ratios). In other words, one has to control only the division step, since multiplication by H is an operator whose norm is no greater than $\|H\|$.

The division step is also rather transparent, its well-posedness being determined by the norm of the inverse Sylvester matrix $\mathfrak{J}_L$ from (25) and the norm of the nonprincipal terms $\|H - L\|$, where L is the principal homogeneous part of H. The group of affine transformations of the complex plane $\mathbb{C}^2$ acts naturally on the space of all Hamiltonians of degree $n+1$ not affecting the critical values of H. By an appropriate transformation of this group, one can always achieve the normalizing condition $\|\mathfrak{J}_L^{-1}\| = 1$ that is a condition on the principal homogeneous part L. The problem on bounding the norm $\|A\|$ is therefore reduced to studying how the magnitude of the nonprincipal coefficients of $H - L$ may affect the configuration of the critical values of H subject to the above normalizing condition on the principal part L. The problem can be explicitly solved for the univariate polynomials, implying an answer in the hyperelliptic case as well [**NY5**]. Moreover, one can show that if all critical values of a bivariate polynomial transversal to infinity coincide, then necessarily the polynomial H must coincide with its principal part, thus being homogeneous, psssibly after a suitable translation in the (x, y)-plane. The inequality between the nonhomogeneity of a bivariate polynomial and the spread of its critical values is still unknown, see [**NY5**] for partial results.

2.3.6. *Preliminary conclusion.* The tangential Hilbert problem for generic Hamiltonians gets reduced to the question about the number of isolated zeros of linear combinations of functions satisfying together a system of linear ordinary differential equations with rational coefficients.

The procedure of derivation of this system is very transparent. In particular, it can be written in the hypergeometric form (28) with explicit bounds on the norms of the corresponding matrices A and B. These bounds in turn imply that when reduced to the Fuchsian form (14), the system will have the residue matrices A_j bounded (as usually, in the sense of the norms) in terms of the inverse distance $\max_{i \neq j} \{|t_i - t_j|^{-1}, |t_i|\}$ between singular points of the system.

When this inverse distance tends to infinity (which corresponds to the confluence of singular points), the resulting bounds on the norms $\|A_j\|$ of the residues in (14) blow up. However, this blowing up is of a very specific nature: the monodromy group of the system with one or several confluent singularities remains the same. In particular, the spectral data of the residues remain bounded.

In the subsequent sections we will find out how far away these conditions are from sufficient conditions allowing for an explicit solution of the tangential Hilbert problem.

3. Quasialgebraicity of function fields

Starting from this moment, we will pursue the same path towards the tangential Hilbert problem, but this time in the opposite direction. Namely, we will establish conditions on Fuchsian systems guaranteeing that their solutions are similar to algebraic functions; in particular, they admit explicit bounds for the number of isolated zeros.

3.1. Function fields and their quasialgebraicity. The main objects of study in this section are function fields obtained by adjoining one or several analytic (in general, multivalued) functions to the field $\mathbb{C}$ of complex numbers (or slightly more generally to that of rational functions $\mathbb{C}(t)$). Such fields admit filtration (grading) by degrees.

The goal is to obtain conditions on the field (in terms of properties of the adjoined functions) guaranteeing that the question on the *global* number of isolated zeros of functions from this field can be *algorithmically* (effectively) solved. An accurate definition will be given at an appropriate moment in Section 3.3, after explaining all technicalities pertinent to the problem. We begin with examples illustrating the goals.

3.1.1. *Algebraic functions.* The field of rational functions $\mathbb{C}(t)$ in one variable t possesses the following obvious but nevertheless remarkable property:

(1) any element $f(t) = p(t)/q(t)$, $p, q \in \mathbb{C}[t]$ from this field has a well defined degree $\deg f = \max(\deg p, \deg q)$ (assuming that the representation is irreducible), and
(2) the number of isolated zeros of f counted with multiplicity on the whole projective line $\mathbb{C}P^1 = \mathbb{C} \cup \{\infty\}$ is equal to $\deg f$.

In other words, there is a direct relationship between the combinatorial complexity of the representation of f in the field (i.e., the number of field operations necessary to produce f from constants and the independent variable t), and its analytic complexity measured by the number of isolated zeros.

This example can be easily generalized by considering fields generated by one or several algebraic functions.

Let $f_i(t)$, $i = 1, \dots, n$ be algebraic functions of one variable, defined implicitly by the polynomial equations

$$P_i(t, x_i) = 0, \quad P_i \in \mathbb{C}[t, x_i], \ \deg P_i = d_i, \ i = 1, \dots, n,$$

with respect to x_i. Consider the ring $\mathbb{C}[f_1, \dots, f_n]$ formed by polynomial combinations of the functions f_i, and the corresponding field of fractions $\mathbb{C}(f_1, \dots, f_n)$. Both consist of analytic multivalued functions ramified over a finite point set on the projective line $\mathbb{C}P^1$, though the number of distinct branches of every function is finite (no more than $d_1 \cdots d_n$).

One can define unambiguously the degree of functions in this ring and the respective field as the degree d of the *minimal* representation

$$f = \sum_{|\alpha| \leqslant d} c_\alpha f^\alpha, \quad \alpha = (\alpha_1, \dots, \alpha_n) \in \mathbb{Z}_+^n, \ c_\alpha \in \mathbb{C}, \ f^\alpha = \prod_1^n f_i^{\alpha_i}.$$

PROPOSITION 3. *The total number of isolated zeros of a polynomial combination* $f(t) = P\big(f_1(t), \dots, f_n(t)\big)$, $f \in \mathbb{C}[f_1, \dots, f_n]$ *of degree* $\leqslant d$ *on all branches of this function does not exceed* $d \cdot d_1 \cdots d_n$.

PROOF. This is an immediate corollary to the Bézout theorem applied to the system of algebraic equations

$$\begin{cases} P_1(t, x_1) = 0, \\ \quad \vdots \\ P_n(t, x_n) = 0, \\ P(x_1, \dots, x_n) = 0. \end{cases}$$

The same proof actually works for a more general case of functions f_i defined by a system of algebraic equations $P_i(t, x_1, \dots, x_n) = 0$, $i = 1, \dots, n$, involving all functions *simultaneously*. □

REMARK. The function $f_0(t) = t$ is clearly algebraic, and if required, we can always assume it being among the collection of the functions f_i, thus avoiding particular cases and awkward notation. This agreement will allow us to assume that all function fields are extensions of the field $\mathbb{C}(t)$.

3.1.2. *Existential quasialgebraicity.* Some parts of the above construction can be reproduced in a completely general context. Let $U \subset \mathbb{C}$ be an open domain (for simplicity assume it to be bounded, connected and simply connected).

Consider an arbitrary collection of n functions $F = \{f_1(t), \dots, f_n(t)\}$ analytic in U. They define the ring $\mathbb{C}[F]$ and the corresponding field $\mathbb{C}(F)$ of functions meromorphic in U. As before, for any function $f \in \mathbb{C}[F]$ one can define its degree as the minimal possible degree of the polynomial $P \in \mathbb{C}[x_1, \dots, x_n]$ realizing the given function $f(t) = P\big(F(t)\big) = P\big(f_1(t), \dots, f_n(t)\big)$ (there can be algebraically dependent functions among the generators). This grading extends naturally for rational combinations from $\mathbb{C}(F)$.

PROPOSITION 4. *For any compact* $K \subset U$ *there exists a counting function* $C = C_K \colon \mathbb{N} \to \mathbb{N}$, *taking only finite values* $C_K(d) < +\infty$ *for any finite* d, *such that the number of isolated zeros of any function* $f \in \mathbb{C}(F)$ *in the compact* K *can be at most* $C_K(d)$:

$$\deg f \leqslant d \implies \#\{t \in K \colon f(t) = 0\} \leqslant C_K(d).$$

PROOF. first we show that under the assumptions of the theorem, the number of isolated zeros of any *linear* combination $f_c(t) = \sum c_i f_i(t)$ with complex constant coefficients $c_1, \dots, c_n \in \mathbb{C}$ is bounded in any compact K uniformly over all such linear combinations. Indeed, without loss of generality we may assume that the functions f_i are linearly independent — this does not affect the supply of all linear combinations. Next, it is sufficient to consider only combinations with coefficients on the unit sphere, satisfying the equality $\sum_j |c_j|^2 = 1$. The functions f_c for such c are all different from the identically zero function, hence each of them has only a finite number of isolated zeros in the compact K (accumulation of roots to the boundary of K is forbidden). Now the standard semicontinuity arguments using compactness of the unit sphere prove that the number of zeros of all f_c is uniformly bounded.

To deal with arbitrary polynomial combinations, we can treat them as linear combinations of *monomials* $f^\alpha(t)$, $|\alpha| \leqslant d$, reducing the general case to the one already studied. □

REMARK. One can easily recognize in this demonstration some minor variations on the theme already exposed in Section 1.1. Of course, Proposition 4 follows from Theorem 2, since the parameters c_α can be considered as varying over the compact sphere. Here we could explicitly avoid dealing with functions vanishing identically for some values of the parameters, considerably simplifying the proof.

3.1.3. *Comparison.* The two above finiteness assertions, Propositions 3 and 4, differ in two important instances:

(1) the bounds on roots of algebraic functions are *global*, i.e., valid on the maximal domain of definition of the functions from the field $\mathbb{C}(F)$, whereas the bounds on the roots of arbitrary analytic functions in general depend on the choice of the compact K and can in many cases blow up as the compact approaches the boundary of the maximal domain U;
(2) the bounds on roots of algebraic functions are given by an explicit formula involving some basic parameters defining the field, whereas the function $C_K(d)$ is totally existential (see below).

The nature of the counting function $C_K(d)$ from Proposition 4 remains totally noneffective. One can easily construct examples of functions (even entire functions) such that the growth of $C_K(d)$ will be arbitrarily fast [**IY3**].

3.1.4. *In search of quasialgebraicity: reappearance of Picard – Vessiot extensions.* Our goal is to provide sufficient conditions on the functions f_i guaranteeing that the corresponding field will be similar to the field obtained by adjoining algebraic functions. This condition, still understood informally until made precise in Section 3.3, will be referred to as *quasialgebraicity* of the function field. The exact definition is postponed since it involves some technical details.

However, even prior to giving any exact formulation, the class of function fields among which one could hope to find nontrivial cases of quasialgebraicity can be substantially restricted.

First, *the generating functions $f_1, \dots, f_n$ must be multivalued* (ramified). Indeed, a single-valued function having at most polar singularities on the projective line $\mathbb{C}P^1$ (recall that we are looking for *global* bounds, hence the functions should be defined globally), must be rational. Any field generated by such functions is a subfield of $\mathbb{C}(t)$ and hence we get nothing new.

On the other hand, if one of the functions f_i has an essential singularity on $\mathbb{C}P^1$, then by classical theorems of complex analysis this function near such a point must take almost all values infinitely many times, hence one can easily construct an *individual* polynomial combination having infinitely many roots accumulating to the essential singularity. This precludes quasialgebraicity, whatever exact meaning may be assigned to it.

Thus any field (or ring, which is almost the same for our purposes) exhibiting nontrivial quasialgebraicity must consist of functions ramified over some finite set $\Sigma \subset \mathbb{C}P^1$. As above, globality means that functions should be analytically continuable along any path avoiding the ramification locus Σ.

The possibility of analytic continuation along paths (and loops) introduces an additional structure, the monodromy group action. Choose arbitrarily a nonsingular point $a \notin \Sigma$. Then any element from the ring $\mathbb{C}[F]$ can be identified with the full analytic continuation of its germ at a. Denote as in Section 2.1 the monodromy operator associated with a loop $\gamma \in \pi_1(\mathbb{C} \setminus \Sigma, a)$ by Δ_γ. Then it would be natural to assume that $\mathbb{C}[F]$ is closed (invariant) under analytic continuation, that is,

$$\Delta_\gamma f_i \in \mathbb{C}[f_1, \dots, f_\mu] \quad \forall \gamma \in \pi_1(\mathbb{C}P^1 \setminus \Sigma, a). \tag{30}$$

Moreover, since the filtration of the ring $\mathbb{C}[F]$ by degrees should be well-defined, it must be preserved under analytic continuation, that is, continuation of a polynomial $f = P(f_1, \dots, f_\mu)$ (of minimal degree among all polynomials representing the given f) along any loop should again be a polynomial of the same degree. In particular, *the $\mathbb{C}$-linear span of the germs $f_1, \dots, f_\mu$ in the space of all analytic germs must be invariant under all monodromy operators.* In other words, there should exist invertible square matrices M_γ such that

$$\Delta_\gamma(f_1, \dots, f_n) = (f_1, \dots, f_n) \cdot M_\gamma.$$

In the same way as in Section 2.1, this implies that the functions $f_1, \dots, f_\mu$ must be solutions of a linear ordinary differential equation with single-valued coefficients on $\mathbb{C}P^1 \setminus \Sigma$. As before, allowing essential singularities of the coefficients would immediately relinquish any control over nonaccumulation of roots of solutions, hence we arrive at the following important conclusion: *To be quasialgebraic, the field $\mathbb{C}(F) = \mathbb{C}(f_1, \dots, f_\mu)$ must be a Picard–Vessiot extension of $\mathbb{C}$ or $\mathbb{C}(t)$ obtained by adjoining solutions of a linear ordinary differential equation with rational coefficients.*

This fact, in particular, implies that adding to the generating tuple of functions $(f_1, \dots, f_n)$ their derivatives of orders $\leqslant n-1$ will make the field $\mathbb{C}(t, f_1, \dots, f_n)$ a *differential field* (closed under differentiation). Without loss of generality we can assume that such a completion was already done and we are dealing with the differential field $\mathbb{C}(X)$ (and the corresponding ring $\mathbb{C}[X]$) obtained by adjoining to $\mathbb{C}$ all entries of a fundamental matrix solution $X(t)$ for a system of first order linear ordinary differential equations with rational coefficient matrix,

$$\dot X(t) = A(t)X(t), \quad A(t) \in \operatorname{Mat}_{\mu\times\mu}(\mathbb{C}(t)). \tag{31}$$

In the particular case we are discussing, one can take X to be the Wronskian matrix of the collection $\{f_i\}$, $X_{ij}(t) = f_j^{(i-1)}(t)$, $i, j = 1, \dots, n$.

3.1.5. *Quasialgebraicity of Fuchsian systems: examples and counterexamples.* As in the case of single-valued functions, a precondition for quasialgebraicity is nonaccumulation of roots of solutions to singular points. Singularities of linear systems (31) can be easily described: they occur only at the poles of the coefficient matrix $A(t)$ or rather at the poles of the matrix-valued differential 1-form $A(t)\,dt$ on the Riemann sphere $\mathbb{C}P^1$. The corresponding classical theory [**H2**] distinguishes between two types of singularities:

- those exhibiting at most polynomial growth of entries of the fundamental matrix $X(t)$ and its inverse $X^{-1}(t)$ and called *regular singularities*, and
- those exhibiting abnormally fast (faster than polynomial) growth of solutions, called *irregular singularities*.

REMARK. In order to measure the growth rate of multivalued functions near a ramification point, they should be restricted to a sector bounded by two rectilinear rays with the vertex at this point. Otherwise one can construct a curve approaching the singular point while spiraling around it in such a way that the growth in terms of the distance to the singularity will be arbitrarily fast even for the most innocent multivalued function $\ln t$.

The dichotomy between regular and irregular singularities is closely related to the dichotomy between poles and essential singularities for single-valued functions. Consider a neighborhood of a singular point $t_* \in \Sigma$, assuming for simplicity that $t_* = 0$. Let M be the monodromy operator associated with a small loop around the origin, so that

$$\Delta_\gamma X(t) = X(t)M, \quad \det M \neq 0. \tag{32}$$

Let $A \in \operatorname{Mat}_{\mu\times\mu}(\mathbb{C})$ be any matrix logarithm of M, so that

$$\exp 2\pi i A = M.$$

Then the multivalued matrix function $t^A = \exp(A \ln t)$ also has the same monodromy,

$$\Delta_\gamma t^A = \exp(A\Delta_\gamma \ln t) = \exp\bigl(A(\ln t + 2\pi i)\bigr) = t^A M = M t^A.$$

Therefore the matrix ratio $H(t) = X(t)t^{-A}$ is single-valued in a small punctured neighborhood of 0. If the origin is a regular singularity, then $H(t)$ has at most a pole and hence by choosing a different valuation of the logarithm and, replacing t^A by t^{kE+A} for sufficiently large natural k, one can make $H(t)$ holomorphic at the origin. By explicit elementary formulas for matrix exponents, one can derive from this a local representation of entries of the matrix $X(t)$ in the form

$$\sum_{k,\lambda} h_{k\lambda}(t)\, t^\lambda \ln^{k-1} t$$

with the exponents λ ranging over the spectrum of A, the natural number k being no greater than the maximal size of Jordan blocks of A and $h_{k\lambda}(t)$ holomorphic at $t = 0$. Notice the remarkable coincidence with (12) and Exercise 1.2: the latter implies that, at least in the situation when all functions $h_{k\lambda}$ and all eigenvalues of A are real, isolated zeros cannot accumulate to the regular singularity at the origin.

If, on the other hand, the singularity is irregular, then at least some entries of the matrix $H(t)$ must exhibit an essential singularity at the origin. In the same way as with single-valued function, one can in this case construct functions from $\mathbb{C}(X)$ that would have infinitely many roots accumulating to the origin.

EXAMPLE 3.1. The linear system

$$\begin{cases} \dot x_1 = t^{-2} x_1, \\ \dot x_2 = 0 \end{cases} \tag{33}$$

generates the field $\mathbb{C}(X)$ containing a function $f(t) = \exp(-1/t) - 1$ whose zeros at the points $t_m = (2\pi i m)^{-1}$, $m = \pm 1, \pm 2, \dots$ accumulate to two essentially singular points at $t = 0$ and $t = \infty$ along the imaginary axis. ◀

Thus occurrence of irregular singularities destroys any hope of achieving quasialgebraicity, and we are left with the class of Picard–Vessiot extensions exhibiting only regular singularities.

There is a simple sufficient condition guaranteeing that a singular point of the linear system (31) is regular. By the Fuchs theorem [**H2**], if $A(t)$ has a *simple pole* (of the first order) at a point t_*, then this point is a regular singularity. Such singularities are called *Fuchsian.*

The converse to the Fuchs theorem is in general not true: there exist regular non-Fuchsian singularities. But a system exhibiting a regular singularity at t_* can be reduced to a system having the simple pole at t_* by a *meromorphic* (locally near t_*) linear transformation $X(t) \mapsto R(t)X(t)$. Thus on the level of *local meromorphic equivalence* there is no difference between Fuchsian and regular singularities.

For a globally defined system exhibiting several regular singularities at finitely many points $\Sigma = \{t_1, \dots, t_d\}$ on the projective line, one can ask whether there exists a globally meromorphic (hence *rational*) linear transformation simultaneously taking all regular singularities into Fuchsian ones. The problem (which constitutes a part of the so called Riemann – Hilbert, or 21st Hilbert problem) turns out to be very delicate, the result depending essentially on the structure of the monodromy group of the system, and not always admitting solution, as shown recently by A. Bolibrukh [**B4, B2**]. However, from the classical result of Plemelj [**P4, F1**] it follows that if an additional singular point is allowed to be created anywhere, then the answer is always positive and a rational matrix function $R(t)$ can be found such that $Y(t) = R(t)X(t)$ satisfies a system of linear ordinary differential equations having only Fuchsian singularities on the whole projective line $\mathbb{C}P^1$. Clearly, if $t \in \mathbb{C}(X)$ (which we may always assume without loss of generality, as noted above), then $\mathbb{C}(X) = \mathbb{C}(Y)$ and hence when discussing quasialgebraicity one can deal with Fuchsian systems only.

A Fuchsian system with d finite singular points $t_1, \dots, t_d$ can always be written in the form

$$\frac{dx}{dt} = A(t)x, \quad A(t) = \sum_{j=1}^{d} \frac{A_j}{t - t_j}, \tag{34}$$

$$x \in \mathbb{C}^n, \quad A_j \in \operatorname{Mat}_{n\times n}(\mathbb{C}), \quad t \in \mathbb{C}, \quad t_1, \dots, t_d \in \mathbb{C},$$

explicitly indicating the corresponding residue matrices A_j (we return to the initial notation X for the dependent variables). Thus the natural problem arises, *When is the Picard – Vessiot extension $\mathbb{C}(X)$ constructed by adjoining all components of the fundamental matrix solution of a Fuchsian system* (34) *quasialgebraic*? Note that this field depends not on the choice of the matrix solution $X(t)$ but rather on its "logarithmic derivative" $A(t) = \dot X(t)X^{-1}(t)$ which is a rational matrix function. The Fuchsian system (34) is determined by its dimension and the collection of algebraic data $\{A_i, t_i,\ i = 1, \dots, d\}$. The bound for the number of zeros should be given in terms of these algebraic data.

EXAMPLE 3.2. The simplest class of Fuchsian systems is that having only two singularities (one simple pole of the matrix $A(t)$ on the whole line $\mathbb{C}P^1$ is impossible since the sum of all the residues, including the one at infinity, must be zero). By a conformal transformation of the independent variable the two points can be placed at $t = 0$ and $t = \infty$. The corresponding system will then take the *Euler form,*

$$\dot X = \frac{A}{t} \cdot X, \quad A \in \operatorname{Mat}_{\mu\times\mu}(\mathbb{C}). \tag{35}$$

The Euler system (35) can be immediately integrated: $X(t) = t^A$. The associated Picard – Vessiot extension has the form that is already familiar:

$$\mathbb{C}(X) = \mathbb{C}(\{t^\lambda \ln^{k-1} t\}_{\lambda,k}), \quad \lambda \in \operatorname{Spec} A, \quad k \leqslant m(\lambda), \tag{36}$$

where $m(\lambda) \geqslant 1$ is the maximal size of Jordan cells corresponding to the eigenvalue λ.

Roots of *real* functions from the field $\mathbb{C}(X)$ cannot accumulate to the origin if all eigenvalues of A are real (Example 1.2). We will see that the "Euler field" for a system with only real spectrum is indeed quasialgebraic.

On the other hand, the field $\mathbb{C}(t^{\sqrt{-1}}, t^{-\sqrt{-1}})$ associated with the 2×2-Euler system with nonreal spectrum $\{\pm\sqrt{-1}\}$ contains the function $\cos \ln t$ that obviously has an infinite number of roots accumulating to $t = 0$.

This suggests that quasialgebraicity of the fields $\mathbb{C}(X)$ should be somehow related to spectral properties of the residue matrices. ◀

3.1.6. *Counting zeros of multivalued functions globally.* Since functions from the Picard – Vessiot extensions are in general multivalued, counting their zeros "on the whole Riemann sphere" should be supplied with a precise meaning.

EXAMPLE 3.3. The function $f(t) = t^\lambda - 1$ for a real irrational λ has infinitely many different roots $t_k = q^k$, $q = \exp \frac{2\pi i}{\lambda}$, $k \in \mathbb{Z}$, lying on different branches of this function.

In order to single out a branch, it is required to choose a simply connected domain $U \subset \mathbb{C} \setminus \Sigma$. Usually this choice is done by cutting the complex plane along segments or rays with endpoints in the singular locus Σ. In the above example, cutting $\mathbb{C}$ along the negative semiaxis yields a simply connected domain U with the following property: *any branch of the function $t^\lambda - 1$ in this domain has at most $\lfloor \lambda^{-1} \rfloor$ roots in it* (the same bound also holds for $-U$). Here and below $\lfloor a \rfloor$ stands for the integer part of a real number a. This explicit bound is global in the sense that U and $-U$ together cover $\mathbb{C} \setminus \Sigma$. ◀

However, the choice of the simply connected domain U can even affect the finiteness of bounds on the number of zeros.

EXAMPLE 3.4. If λ is a nonreal number, then the roots fall on the logarithmic spiral (since $|q| \neq 1$ in this case). If U is chosen by cutting $\mathbb{C}$ along the positive or negative semiaxis, then the number of isolated roots on any branch will still be bounded by $\lfloor (\operatorname{Re} \lambda)^{-1} \rfloor < +\infty$ for $\lambda \notin i\mathbb{R}$.

On the other hand, one can choose a simply connected spiral domain containing the origin on the boundary such that it simultaneously contains an infinite number of the points $t_k = q^k$.

Though the spiral cuts are not algebraic (not even analytic) curves, this example can be easily modified to construct *polygonal* simply connected domains containing as many roots of the function $t^\lambda - 1$ as necessary, provided that the number of sides of the polygon can be arbitrary. ◀

These examples motivate the following definition.

DEFINITION 6. An analytic multivalued function $f : \mathbb{C}P^1 \to \mathbb{C}$ ramified over a finite set $\Sigma \subset \mathbb{C}P^1$ is said to admit a *global bound on the number of isolated zeros* if there exists a natural number $N < +\infty$ such that the number of isolated zeros of any branch of f in any *open rectilinear triangle* $T \subset \mathbb{C} \setminus \Sigma$ has no more than N isolated zeros, the bound being uniform over all such triangles.

REMARK. Having this property, one can place an explicit upper bound on the number of isolated roots of f in any simply connected domain U bounded by finitely many algebraic curves of known degrees. The number of curves, their degrees and the number of ramification points will explicitly enter the answer together with the number N.

Indeed, the domain $\mathbb{C} \setminus \Sigma$ can be triangulated into finitely many triangles as above. The number of simply connected components of any intersection $U \cap T$ with each triangle of the triangulation can be easily majorized by the Bézout theorem in terms of the above data. The total number of such components should be multiplied by N to construct the required upper bound.

EXAMPLE 3.5. The function $t^\lambda - 1$ admits a global upper bound for the number of isolated roots for all λ with $\operatorname{Re}\lambda \neq 0$. For $\lambda = \sqrt{-1}$ any real interval $(0, a)$, $a > 0$, contains infinitely many roots of this function. ◀

REMARK. In practice, however, we will always choose a convenient system of rectilinear cuts as in textbooks on complex variables: if, for example, an upper bound for the number of zeros is known in both $U = \mathbb{C} \setminus \mathbb{R}_+$ and $-U = \mathbb{C} \setminus \mathbb{R}_-$, then the number of zeros in any triangle not containing the origin does not exceed the maximum of these two bounds, since such a triangle cannot intersect both the positive and negative semiaxes simultaneously.

3.2. Digression on computability.

3.2.1. *Hierarchy of computability.* The discussion in the previous section addressed the issue of *globality* of bounds on the number of zeros of multivalued functions. Now we pass to a brief discussion of *computability* of these bounds. This would lead us again to functions defined by differential equations with polynomial right-hand sides.

The bounds for functions of the field $\mathbb{C}(f_1, \dots, f_\mu)$, besides being dependent on the degree d of these functions, should obviously depend on the field itself, that is, ultimately, on the choice of the generators f_i. While computability of functions of natural arguments is a well developed area (several notions of computability can be distinguished, see below), dependence on functional parameters is definitely outside the scope of any theory (e.g., the number of roots of a function is not a computable characteristic *per se*).

An intermediate position is occupied by "computable" functions of one or several real or complex variables. Here we take the most restrictive attitude, legitimizing only *polynomial functions*, the absolute value $|\cdot|$ of a real or complex variable, and the integer part $\lfloor \cdot \rfloor$ of a real variable.

Returning to functions of one or several natural number arguments, one can classify them in an "increasing order of computability" into:

(1) (general) recursive functions, whose values can be computed for any combination of arguments in a finite time by some deterministic algorithm, Turing machine etc.;
(2) primitive recursive functions, that can be defined by one or several iterated inductive rules of the form

$$\phi(k+1, m) = \Phi\big(k, m, \phi(k, m)\big), \quad k \in \mathbb{N}, \ m \in \mathbb{N}^n, \tag{37}$$

assuming that the functions Φ and $\phi_1 = \phi(1, \cdot)$ are already known;
(3) elementary functions, given by finite compositions of powers, logarithms and exponents.

EXAMPLE 3.6. [tower functions] Consider the function $\tau(k,n)$ of two integer arguments, defined by the recursive rule

$$\tau(0,k)=k, \quad \tau(n+1,k)=2^{\tau(n,k)}. \tag{38}$$

These rules define *tower functions* (iterated exponents): $\tau(n,\cdot)$ is a tower of height n and for each particular n is an elementary function of the second argument k:

$$\tau(1,k)=2^k, \quad \tau(2,k)=2^{2^k}, \quad \tau(3,k)=2^{2^{2^k}}, \quad \dots$$

However, as a function of the first argument, it is not elementary. ◀

EXAMPLE 3.7. [Ackermann generalized exponential] Consider the function $A(z,x,y)$ of three natural arguments defined by the recursive rules and initial conditions as follows,

$$\begin{gathered} A(z+1,x,y+1)=A\big(z,x,A(z+1,x,y)\big), \\ A(0,x,y)=y+1, \quad A(1,x,0)=x, \quad A(2,x,0)=0, \\ A(z,x,0)=0, \quad \forall z\geqslant 2. \end{gathered} \tag{39}$$

These rules define each function $\phi=A(z+1,\cdot,\cdot)$ for any particular z unambiguously, provided that the function $\Phi=A(z,\cdot,\cdot)$ is already defined, by the *simple* recursive rule $\phi(x,y+1)=\Phi\big(x,\phi(x,y)\big)$. In other words, the set of conditions defines A for all nonnegative combinations of arguments.

One can easily check that

$$A(1,x,y)=x+1+\cdots+1=x+y, \qquad A(2,x,y)=x+x+\cdots+x=xy,$$

$$A(3,x,y)=xx\cdots x=x^y, \qquad A(4,x,y)=\underbrace{x^{x^{\cdot^{\cdot^{\cdot^{x}}}}}}_{y \text{ times}},$$

etc. This suggests that as a function of the first argument, A is not primitive recursive. The reason is that, unlike in the simple recursive rule (37), the function A is applied to itself (this does not prove that A is not primitive recursive, but the fact remains true and can be rigorously demonstrated). It can also be shown that as a function of the first argument the Ackermann generalized exponential grows faster than any primitive recursive function. ◀

3.2.2. *Transcendental functions defined by algebraic data.* The above brief discussion suggests that in order to speak about computable bounds depending on several analytic functions as parameters, these analytic functions must themselves be defined in terms of finitely many integer, real or complex parameters.

The simplest examples of such functions are polynomials (their coefficients play the role of the parameters) and, slightly more generally, algebraic functions. However, this example does not allow one to produce nontrivial quasialgebraic fields.

Differentiation of algebraic functions leaves them algebraic. On the contrary, integration (taking primitives) in general results in transcendental functions. Another possibility of constructing transcendental functions from algebraic ones is exponentiation. However, both primitives and exponentials are only particular cases of *functions defined by ordinary differential equations with polynomial right hand sides* (e.g., $x(t)=\exp\int f(t)\,dt$ is a solution to the differential equation $\dot x=f(t)x$). This looks like the most general mechanism transforming algebraic input data into transcendental output functions.

Thus we arrive at the following natural conclusion: in order to speak about quasialgebraicity of the function fields $\mathbb{C}(f_1,\dots,f_\mu)$, the generating functions $f_i(t)$ must be defined by polynomial ordinary differential equations or systems of such equations. Then by computability of any bound we would always assume that this bound can be expressed in terms of the (real or complex) coefficients of these equations and/or integer parameters (dimensions, degrees, etc), so that as functions of real or complex parameters these bounds are polynomial, while being elementary (or at worst primitive recursive) functions of the remaining integer variables.

EXAMPLE 3.8. (principal). Suppose that the functions $f_i(t)$, $i = 1,\dots,\mu$, together satisfy a system of polynomial ordinary differential equations of the form

$$\dot{x}_i = \sum_{|\alpha|\leqslant m} c_{i\alpha}x^\alpha, \quad i = 1,\dots,\mu, \ \alpha = (\alpha_1,\dots,\alpha_\mu) \in \mathbb{Z}_+^\mu,$$

with complex coefficients $c_{i\alpha} \in \mathbb{C}$. Then the expressions

$$C_1 = (\max_{i,\alpha} |c_{i\alpha}|)^{N(\mu,m)}, \quad C_2 = N(\mu,m)\cdot\sum_{i,\alpha}|c_{i\alpha}|$$

are both computable characteristics of the collection $\{f_1,\dots,f_\mu\}$ provided that N is an elementary or at worst primitive recursive function of two integer arguments m (the degree) and μ (the dimension of the system). ◀

3.2.3. *Restricted computable bounds.* Explicit computability does not imply (neither does it assume) globality of the bounds. In the same way as the parameters defining the functions f_i may enter the answer, sometimes numerical characteristics of the domain (in our settings, always a polygon $U \subset \mathbb{C}\setminus\Sigma$) may enter into the expressions.

Consider the Picard – Vessiot extension field $\mathbb{C}(X)$ associated with the Fuchsian system (34), and let $U \subset \mathbb{C}$ be a domain (usually polygonal) possibly containing some of the singular points inside.

DEFINITION 7. We say that the field $\mathbb{C}(X)$ is *quasialgebraic in a polygonal domain* U if the number of zeros of any function $f \in \mathbb{C}(X)$ can be bounded in any triangle $T \subset U\setminus\Sigma$ uniformly on all such triangles, in terms of $\deg f$, the parameters of the field (n, d and the residual norm $\max_j \|A_j\|$ and the geometric characteristics of U.

Below the dependence on U is through the distance between the boundary ∂U and the singular locus Σ of the system.

REMARK. Since the independent variable ranges over the Riemann sphere $\mathbb{C}P^1$ and the point $t=\infty$ may well belong to the singular locus, the distance from ∂U to Σ should be defined in such cases as the minimum of the above distance to the finite part of Σ and the number $\inf_{t\in\partial U}|t^{-1}|$ measuring the "distance from ∂U to infinity".

3.3. Quasialgebraicity and uniform quasialgebraicity.

3.3.1. *Quasialgebraicity of Picard – Vessiot extensions: accurate formulation of the problem.* Let $X(t)$ be a fundamental matrix solution of a Fuchsian system of n linear ordinary differential equations (34), and $\mathbb{C}(X)$ (resp., $\mathbb{C}[X]$) the field obtained by adjoining all entries of this matrix to the field $\mathbb{C}$ (resp., the ring of all polynomial combinations of these entries). As was already noted, we always assume that $\mathbb{C}(X)$ contains the subfield of rational functions $\mathbb{C}(t)$.

DEFINITION 8. The field $\mathbb{C}(X)$ is called *quasialgebraic* if the number of isolated roots of any function $f \in \mathbb{C}(X)$ of degree k in this field, in any triangle T free from singular points t_j of the Fuchsian system (34), is bounded by a number depending only on:

(1) the degree $k = \deg_{\mathbb{C}(X)} f$;
(2) the dimension n and the number d of (finite) singular points (as an elementary or at worst primitive recursive function);
(3) the entries of the residue matrices $A_1, \dots, A_d$ (in a rational way);
(4) on the coordinates of the singularities $t_1, \dots, t_d$ (also in a rational way).

In fact, in all cases when quasialgebraicity of the field will be established, the bounds would depend on the complex (matrix) parameters in a very simple way, via the *residual norm* of the rational matrix function $A(t)$ represented as in (34),

$$R\big(A(\cdot)\big) = \max_{j=1,\dots,d} \|A_j\|. \tag{40}$$

In a similar manner, dependence of the bounds on the position of the singularities t_j will be expressed via the (inverse) *spread* of these points,

$$\rho(\Sigma) = \max_{i \neq j} \{|t_i - t_j|^{-1},\ |t_i|\}. \tag{41}$$

(this number is large only when some of the singular points approach each other).

3.3.2. *Uniform quasialgebraicity.* According to the above definition, the field is quasialgebraic if the simple algebraic data (parameters) defining it can be used to produce an explicit bound on the number of zeros of any function from this field.

Some of these parameters are always relevant. There is no question that the dimension and the degree must necessarily enter any bound on zeros. It requires minimal effort to see that other parameters can also affect the answer.

EXAMPLE 3.9. The Euler system

$$\dot x_1 = a\,t^{-1} x_1,\ a \in \mathbb{N}, \quad \dot x_2 = 0,$$

defines a field containing the function $f(t) = t^a - 1$ that is of degree *one* in this field (though, as a polynomial, it has degree a). The number of isolated zeros of the function f in a triangle can be as large as $\lfloor a/2 \rfloor$. ◄

This example shows that the magnitude of eigenvalues of the residue matrices clearly affects the number of zeros.

It is not very difficult (see the subsequent sections) to construct upper bounds (global or not) that would involve *rational* expressions of the parameters (in this case, entries of the residue matrices) that have poles, making the bounds blow up for certain combinations of parameters. On the other hand, there are no visible reasons for the appearance of the infinite number of zeros for these values of the parameters. It requires considerable effort to show that in terms of the residue matrices the bounds can be given by *polynomial* expressions, which is equivalent to expressing them in terms of the residual norm (40) as above. Finally, it is the most difficult part to prove that, under certain additional but rather natural assumptions, the bounds can be given *uniformly over all configurations of singular points.*

3.3.3. *Notes, remarks.* We conclude this highly informal discussion by several remarks, also of a very general nature.

As stated in Section 1.4, the ultimate goal of the theory is to establish a constructive bound for the number of zeros of Abelian integrals (tangential Hilbert problem). These integrals were shown to belong to certain Picard – Vessiot extensions. Application of the methods explained in subsequent sections requires repeated algebraic manipulations with the integrals, therefore bringing into play the whole field of functions. Moreover, in the final stage of the construction not one but rather several Picard – Vessiot fields are considered simultaneously, while carrying out induction on the number of ramification points. This explains why the existence of upper bounds on the number of zeros (quasialgebraicity) is defined as a property of the corresponding function fields (rather than elements constituting these fields).

There was also a special reason for choosing the definition of quasialgebraicity without attempting to specify explicitly the "counting function" measuring the number of isolated zeros. This was done primarily because the bounds that one can obtain in this way are *enormously* excessive. Several nested inductive constructions immediately produce tower-like bounds even from the modest exponential contributions at each inductive step.

Finally, we would like to note that the Fuchsian representation is not the only one possible. Actually, the Picard – Fuchs system of differential equations for Abelian integrals is written in the hypergeometric form

$$(tE + A)\dot{X}(t) = BX(t)$$

determined by two constant matrices $A, B \in \operatorname{Mat}_{n\times n}(\mathbb{C})$, and it would be natural to require that quasialgebraicity be expressed in terms of the norms $\|A\|$, $\|B\|$ rather than in terms of the respective residues after transforming to the Fuchsian form. This type of bound is not proved (if it were, this would imply a constructive solution of the tangential Hilbert problem, as explained in Section 2.3.3).

4. Quantitative theory of differential equations

4.1. Bounded meandering principle: explicit bounds on zeros of functions defined by ordinary differential equations. The preceding section contained motivations for introducing the notion of quasialgebraicity. In this section we survey some results of constructive (though not always global) nature, bounding the number of isolated zeros of functions defined by differential equations.

4.1.1. *Linear n^{th} order equations with bounded analytic coefficients.* The basis for all other considerations is a classical theorem by de la Vallée Poussin concerning solutions of linear ordinary differential equations with bounded coefficients, not necessarily polynomial or even analytic. It gives a sufficient condition for a linear ordinary differential equation of order n with *real bounded coefficients* guaranteeing the absence of solutions with more than $n-1$ isolated roots on a given real interval $I \subset \mathbb{R}$. Such equations are called *disconjugate* on I. In order to stress the difference from the complex case, we denote the independent variable by s, and consider a linear equation

$$y^{(n)} + a_1(s)\, y^{(n-1)} + \cdots + a_{n-2}(s)\, y'' + a_{n-1}(s)y' + a_n(s)y = 0 \tag{42}$$

on the real interval $I = [s_0, s_1]$ of length r with real bounded coefficients: $|a_k(s)| < c_k < \infty$ for all $s \in I$.

LEMMA 8 ([**dlVP**]). *If*

$$\sum_{k=1}^{n} \frac{c_k\, r^k}{k!} < 1, \tag{43}$$

then any C^n-smooth function $f(s)$ satisfying a linear equation (42) *can have at most $n-1$ isolated roots on I, counted with multiplicities.*

This result can be seen as a generalization of the Sturm nonoscillation theorem for equations of order greater than 2. The inequality can be slightly improved, see [**L**].

This (simple) statement implies a number of corollaries. First, subdividing any interval into sufficiently short segments satisfying (43), and adding together the bounds for each segment, one can obtain an explicit bound on zeros valid on any real interval where the coefficients $a_i(s)$ are explicitly bounded. Next, one can consider equations of the same form (42) with *complex-valued* coefficients and solutions. There is no sense in counting zeros of such solutions (since in general there are no roots), instead an interesting question is to estimate their *topological index*, the variation of the argument $\operatorname{Arg} f(s)$ between the endpoints of the real interval. By a simple modification of the method used in the proof of Lemma 8 one can obtain an upper bound for this topological index, again in terms of the magnitude (uniform upper bounds for the absolute value) of coefficients of the equation (42), see [**Y2**], as follows.

LEMMA 9. *The variation of the argument of any solution $f(t)$ of the equation* (42) *with complex-valued coefficients all bounded by C in the absolute value along the segment I of length $|I| = s_1 - s_0$ is no greater than*

$$\pi(n+1)(1+3C|I|).$$

If the equation has *analytic* coefficients $a_i \in \mathcal{O}(U)$ defined in some domain $U \subset \mathbb{C}$, then the number of complex isolated zeros of any solution of (42) inside any polygonal domain can be majorized using the above described bounds for the topological index and the classical argument principle. As a result, the following inequality for the number of isolated zeros can be obtained (we return to the initial notation t for the complex independent variable).

THEOREM 7 ([**Y2**]). *Suppose $f_1(t), \dots, f_n(t)$ solve*

$$\begin{aligned} y^{(n)} + a_1(t)y^{(n-1)} + \cdots + a_{n-1}(t)y' + a_n(t)y = 0, \quad t \in U,\\ a_i \in \mathcal{O}(U),\ |a_i(t)| \leqslant R, \quad i = 1, \dots, n. \end{aligned} \tag{44}$$

Then for any triangle $T \subset U$ of perimeter ℓ any linear combination $f = \sum_1^n c_i f_i$ with complex constant coefficients c_i has no more than

$$\tfrac{3}{2}(n+1)(1+\ell R) \tag{45}$$

isolated zeros in T.

REMARK. Analyticity of solutions of the equation (44) follows from analyticity of its coefficients. However, the theorem apparently can be modified to cover also equations with meromorphic coefficients, provided that the solutions are analytic in U (the so called apparent singularities).

4.1.2. *Systems of linear equations: Novikov's counterexample.* Motivated by potential applications to Abelian integrals and quasialgebraic fields, we would like to find generalizations of Lemma 8 or Theorem 7 for systems of linear equations, in the hope of majorizing the number of isolated roots of any linear combination of coordinates $c_1x_1(s) + \cdots + c_nx_n(s)$ for a system

$$\dot{x}(s) = A(s)x(s), \quad s \in I \subset \mathbb{R}, \quad \forall s \in I \quad \|A(s)\| \leqslant R,$$

in terms of $\ell = |I|$ and R. The following counterexample due to D. Novikov [**N1**], shows that this is impossible, both for real and for complex systems.

EXAMPLE 4.1. Let $t_1, \dots, t_d$ be a collection of different numbers (real or complex) from a real interval I or a triangle T (consider only the last case for simplicity). Consider the polynomial $a(t) = \varepsilon(t-t_1)\cdots(t-t_d)$, where the leading coefficient ε is chosen so small that $|a(t)| + |\dot{a}(t) + a^2(t)| < 1$ for any $t \in T$.

The solution $\phi_1 = \exp(\int a(t)dt)$ of the linear differential equation of first order $\dot{x}_1 = a(t)x_1$ has no zeros at all. However, the derivative $\phi_2 = \dot{\phi}_1 = a(t)\phi_1$ has the same zeros as $a(t)$ and satisfies the equation $\dot{\phi}_2 = (\dot{a} + a^2)\phi_1$.

Together the pair (ϕ_1, ϕ_2) satisfies the linear 2×2-system

$$\dot{x}_1 = ax_1, \quad \dot{x}_2 = (\dot{a} + a^2)x_1, \quad a(t) = \varepsilon\prod_1^d(t-t_j), \tag{46}$$

whose coefficients are bounded by 1 everywhere in T by the choice of ε. However, the second component has the specified number d of isolated zeros there, where d can be arbitrarily large. ◀

This example suggests that a direct generalization of Theorem 7 for *systems* of linear equations is impossible: no *pointwise* upper bound on the norm of the coefficient matrix $A(t)$ can ensure uniform upper bounds on oscillation (the number of isolated zeros) of components of the solutions. Any reasonable statement is possible only for systems with rational or even polynomial coefficient matrices

$$A(t) = \sum_{k=0}^d A_k t^k, \quad A_k \in \operatorname{Mat}_{n\times n}(\mathbb{C}), \tag{47}$$

and the magnitude of the matrix coefficients, e.g., the total norm $R = \sum_k \|A_k\|$ must necessarily enter into the upper bounds for the number of zeros of solutions.

4.1.3. *Reduction of a system to an equation: obstructions.* The above example is not very surprising since the standard procedure of reducing a linear $n \times n$-system

$$\dot{x} = A(t)x, \quad x = (x_1, \dots, x_n) \in \mathbb{C}^n, \ t \in U, \tag{48}$$

with holomorphic matrix coefficients $A(t) \in \operatorname{Mat}_{n\times n}\big(\mathcal{O}(U)\big)$ to one nth order linear equation is discontinuous with respect to the parameters. Indeed, without loss of generality one can assume that it is the number of zeros of the first component $y = x_1$ in the domain $U \subset \mathbb{C}$ that interests us. Differentiating x_1 with respect to t, in accordance with the system, we see that the derivatives are linear combinations:

$$y^{(k)}(t) = \boldsymbol{q}_k(t) \cdot x(t), \quad k = 1, \dots, n, \dots, \tag{49}$$

where the analytic (co)vector functions $\boldsymbol{q}_k(t) = \big(q_{k,1}(t), \dots, q_{k,n}(t)\big) \in \mathcal{O}(U) \otimes \mathbb{C}^{n*}$ are determined by the recurrence rule

$$\boldsymbol{q}_0(t) = (1, 0, \dots, 0), \quad \boldsymbol{q}_{k+1}(t) = \dot{\boldsymbol{q}}_k(t) + \boldsymbol{q}_k(t) \cdot A(t) \tag{50}$$

and do not depend on the choice of the trajectory $x(t)$. (We use the right multiplication to stress that $\boldsymbol{q}_k(t)$ are row vector functions). Consider the field $\Bbbk = \mathcal{M}(U)$ of functions meromorphic in U and the linear n-space $\Bbbk^n$ over this field.

The functions $\boldsymbol{q}_k(\cdot)$ are vectors in this space. Consider the linear subspaces

$$\{0\} \subset L_0 \subseteq L_1 \subseteq \cdots \subseteq L_{n-1} \subseteq L_n \subseteq \cdots \subseteq \Bbbk^n, \tag{51}$$

each L_k being spanned by $\boldsymbol{q}_0, \boldsymbol{q}_1, \ldots, \boldsymbol{q}_k$ over $\Bbbk$. If all inclusions are strict, the dimensions of L_k over $\Bbbk$ must strictly increase, while being bounded by $n = \dim_{\Bbbk} \Bbbk^n$. Hence the ascending chain of subspaces (51) must stabilize no later than after n steps, i.e., for some $\mu \leqslant n$ the inclusion must be nonstrict, $\boldsymbol{q}_\mu \in L_{\mu-1}$, which implies a linear identity over $\Bbbk = \mathcal{M}(U)$,

$$\boldsymbol{q}_\mu + a_1(t)\boldsymbol{q}_{\mu-1} + \cdots + a_{\mu-1}(t)\boldsymbol{q}_1 + a_\mu(t)\boldsymbol{q}_0 = 0, \quad a_i \in \Bbbk = \mathcal{M}(U). \tag{52}$$

This implies that $y = \boldsymbol{q}_0 \cdot x$ satisfies the equation (44).

Though the recurrence formulas generating the vectors $\boldsymbol{q}_1, \boldsymbol{q}_2, \ldots$ are explicit, it is impossible to control the magnitude of the coefficients $a_i(\cdot)$, whatever that may mean, from this construction. This can be seen in a simpler setting when the function field $\Bbbk$ is replaced by $\mathbb{C}$.

EXAMPLE 4.2. Consider a sequence of vectors $q_0, q_1, \ldots,$ in the n-dimensional space over $\mathbb{R}$ or $\mathbb{C}$ that grows at most exponentially in the sense of the norm: $\|q_k\| \leqslant c^k$, $c > 0$.

For the same dimensionality reasons as above, no later than at the nth step a linear dependence must occur, allowing one to express some q_k as a linear combination of the preceding vectors $q_0, \ldots, q_{k-1}$,

$$-q_k + a_1 q_{k-1} + \cdots + a_k q_0 = 0, \quad a_i \in \mathbb{C}, \tag{53}$$

similarly to (52). However, the coefficients of this dependence are out of control and can be arbitrarily large. To see this, consider the situation when the angle between q_{k-1} and L_{k-2} is very small but nonzero so that $L_{k-2} \subsetneq L_{k-1}$, while q_k belongs to L_{k-1} and is orthogonal to L_{k-2} in L_{k-1}. Even assuming some *lower* bounds on $\|q_k\|$ will not help to improve the situation in this case. ◀

The situation becomes completely different under either of the following two additional assumptions.

EXAMPLE 4.3. Assume that all the vectors q_k actually belong to a lattice $\mathbb{Z}^n \subset \mathbb{C}^n$ and their norms are bounded from above as in the preceding example. Then the coefficients of the linear combination (52) can be explicitly bounded from above: indeed, if q_k is a linear combination of $q_0, \ldots, q_{k-1}$, then the coefficients a_i in (53) can be found by solving a system of linear nonhomogeneous algebraic equations with integer matrix of coefficients and integer free terms. By the Cramer rule, the solutions of this system can be obtained (after elimination of redundant equations and assuming that solutions indeed exist) as ratios of appropriate minors (determinants of some square submatrices). These minors are integers, explicitly bounded from above by virtue of the assumptions on the norms $\|q_i\|$. Hence in each ratio the numerator is bounded from above, whereas the denominator is no smaller than 1 in absolute value. This clearly implies an upper bound on all the coefficients a_i. ◀

EXAMPLE 4.4. Assume that the vectors $q_k \in \mathbb{C}^n$ are obtained by iterations of a linear map $P\colon \mathbb{C}^n \to \mathbb{C}^n$. Then the bounds on the norms will be automatically satisfied with $c = \|P\|$. The *first* linear combination between the vectors can still have very large coefficients, exactly as in Example 4.2. However, instead of looking for the first combination, we may continue until step number n. Since any operator P is a matrix root of its characteristic polynomial $\chi(z) = z^n + a_1 z^{n-1} + \cdots + a_{n-1}z + a_n$, the linear combination of the form (53) in this case can be obtained via the identity $\chi(P)q_0 = 0$. The coefficients a_i of the characteristic polynomial admit an upper bound in terms of the eigenvalues of P, each of which is no greater than the norm $\|P\|$. Finally, we conclude that a linear combination (52) in this case (iterations of a linear map P in the finite-dimensional space $\mathbb{C}^n$) can be constructed with coefficients a_i explicitly bounded in terms of $\|P\|$ and n. ◀

4.1.4. *Reduction of a system to an equation: commutative algebra versus linear algebra.* The operator $\boldsymbol{q}(t) \mapsto \dot{\boldsymbol{q}}(t) + \boldsymbol{q}(t) \cdot A(t)$ from the n-dimensional $\mathbb{C}(t)$-linear space $\mathbb{C}(t) \otimes \mathbb{C}^n$ into itself, describing the iterations (50), is *not* linear over the field $\mathbb{C}(t)$, since the derivative d/dt is not a $\mathbb{C}(t)$-linear operator. On the other hand, considered over the field $\mathbb{C}$, the space $\mathbb{C}[t] \otimes \mathbb{C}^n$ of polynomial vector-functions is *not* finite-dimensional (and indeed the degrees of iterations (50) grow to infinity unless the coefficients of the system are constant). Thus neither of the above methods can work.

However, a solution can be found in terms of commutative algebra. It allows one to treat in a similar way both linear and nonlinear systems. Before proceeding further, we give a useful technical definition.

DEFINITION 9. The *norm* of a polynomial $p \in \mathbb{C}[x_1, \dots, x_n]$ is the sum of the absolute values of its coefficients:

$$\Big\|\sum\nolimits_\alpha c_\alpha x^\alpha\Big\| = \sum\nolimits_\alpha |c_\alpha|. \tag{54}$$

One can easily verify that in addition to the usual triangle inequality, this norm is multiplicative, $\|pq\| \leqslant \|p\|\,\|q\|$ for any two polynomials. The norm of a derivative $\|\partial_i p\|$ (in any variable) can be easily bounded, $\|\partial_i p\| \leqslant \deg p\, \|p\|$.

Generalization of the construction from Section 4.1.3 is very simple.

EXAMPLE 4.5. (basic). Consider the linear system (48) with a *polynomial* matrix of coefficients $A(t)$ as in (47). The rule (50) defines a sequence of *polynomials* $q_k(t,x) = \boldsymbol{q}_k(t) \cdot x \in \mathbb{C}[t,x]$ that are always linear in the variables $x = (x_1, \dots, x_n)$: the recursion rules allow one to estimate the degrees $\deg_t q_k(t,x)$ and, if necessary, their norms.

Instead of the linear subspaces L_k over the field $\Bbbk = \mathbb{C}(t)$ in this case, consider the *polynomial ideals* $I_k = (q_0, \dots, q_k) \subset \mathbb{C}[t,x]$. The ascending chain

$$\{0\} \subset I_0 \subseteq I_1 \subseteq \cdots \subseteq I_n \subseteq \cdots \subseteq \mathbb{C}[t,x] \tag{55}$$

of these ideals must eventually stabilize in the sense that some inclusion $I_{\ell-1} \subset I_\ell$ becomes nonstrict (an equality). This follows from the fundamental fact that the polynomial ring $\mathbb{C}[t,x]$ is Noetherian and *any* ascending polynomial chain in it eventually stabilizes.

The above stabilization condition means that $\pm q_\ell \in I_{\ell-1}$, hence for appropriate polynomial coefficients $h_1, \dots, h_\ell \in \mathbb{C}[t, x]$,

$$q_\ell + \sum_{i=1}^{\ell} h_i q_{\ell-i} = 0. \tag{56}$$

A priori, the polynomial coefficients h_i can depend on x in a nontrivial way. However, since all polynomials q_k are linear in x (homogeneous), one can truncate the identity (56) retaining only constant terms of h_i (of degree 0 in x) and construct a new identity of exactly the same form (56) but with coefficients $a_i(t) = h_i(t, 0)$ from the *univariate* polynomial ring $\mathbb{C}[t]$,

$$q_\ell + \sum_{i=1}^{\ell} a_i q_{\ell-i} = 0, \quad a_i \in \mathbb{C}[t],\ i = 1, \dots, \ell. \tag{57}$$

The identity (57) obviously means that the function $q_0(t, x) = x_1$ satisfies the linear ordinary differential equation

$$y^{(\ell)} + \sum_{i=1}^{\ell} a_i(t)\, y^{(\ell-i)} = 0, \quad a_i(t) = h_i(t, 0) \in \mathbb{C}[t],\ i = 1, \dots, \ell. \tag{58}$$

In order to apply the results form Section 4.1.1, one has to estimate the absolute values $|a_i(t)|$ from above. In any disk of known radius $\{|t| < R\}$ this can be done if the decomposition (56) is explicitly known, in particular, if the following parameters are explicitly bounded from above:

- the length ℓ of the chain (equal to the order of the resulting differential equation);
- the degrees of the coefficients h_i in t;
- the norms $\|h_i\|$, $i = 1, \dots, \ell$. ◀

EXAMPLE 4.6. The construction involving chains of ideals is not very degree-specific. Consider a system of *polynomial* ordinary differential equations,

$$\dot{x}_i = P_i(t, x), \quad P_i \in \mathbb{C}[t, x_1, \dots, x_n],\ i = 1, \dots, n. \tag{59}$$

Let f be a polynomial combination $f(t) = Q\big(t, x_1(t), \dots, x_n(t)\big)$ for some $Q = Q(t, x) \in \mathbb{C}[t, x]$. Consider the infinite sequence of polynomials $q_0, q_1, \dots, q_n, \dots \in \mathbb{C}[t, x]$ formed by iterations of the Lie derivative, i.e., the recursive rule

$$q_0 = Q, \quad q_{k+1} = \frac{\partial q_k}{\partial t} + \sum_{i=1}^{n} \frac{\partial q_k}{\partial x_i} \cdot P_i. \tag{60}$$

Let $I_k = (q_0, q_1, \dots, q_k) \subset \mathbb{C}[t, x]$ be the polynomial ideals generated by the first $k + 1$ polynomials from this sequence in the polynomial ring $\mathbb{C}[t, x]$. They obviously form an ascending chain, $I_k \subset I_{k+1}$. Since the ring $\mathbb{C}[t, x]$ is Noetherian, this chain must stabilize and hence for some natural number ℓ,

$$q_\ell + h_1 q_{\ell-1} + \dots + h_{\ell-1} q_1 + h_\ell q_0 = 0, \quad h_i \in \mathbb{C}[t, x]. \tag{61}$$

As in the linear case, this identity implies a polynomial relationship between the unknown function $y = Q\big(t, x(t)\big)$ and its derivatives up to order ℓ. Unlike the linear case, this time the coefficients h_i may depend on x explicitly, so that in addition to the data described in the previous example one needs upper bounds for $|x_i(t)|$ in U, which is a nonalgebraic piece of information. However, in many cases this information can be easily achieved (or even *a priori* known) and the problem reduces to getting the same information as in the linear case, namely: the length ℓ of the ascending chain of polynomial ideals, the degrees and the norms of the polynomials h_i appearing in the representation (61). ◀

4.1.5. *Generalizations and improvements.* It is not clear from the very beginning how replacing ascending chains of linear subspaces by ascending chains of polynomial ideals may resolve the problems related to unboundedness of the coefficients of the decompositions (52) and (61) respectively, see Example 4.2. We explain it now.

The first advantage of the suggested approach allows one to treat systems depending polynomially on additional parameters, without the risk of producing bounds that blow up for certain values of the parameters. Actually, the difference between parameters and phase variables disappears almost completely.

EXAMPLE 4.7. Assume that the linear system (31) with a polynomial matrix $A(t)$ as in (47) depends on additional parameters $\lambda = (\lambda_1, \dots, \lambda_p)$ in a polynomial way: $A \in \operatorname{Mat}_{n\times n}(\mathbb{C}[t,\lambda])$. Then one can consider the chains of ideals in the bigger ring $\mathbb{C}[t,\lambda]$ and only minor notation changes are necessary to construct a linear ordinary differential equation (58) whose coefficients will be in fact polynomial in t *and* λ (note that the resulting differential equation is always monic: its leading coefficient of the principal derivative is 1).

This polynomiality eliminates the danger that for some values of the parameters the coefficients of the derived equation (58) will blow up (which was earlier the case). ◀

Another advantage appears as a generalization of Example 4.3.

EXAMPLE 4.8. Assume that the coefficient matrix $A(t)$ of the system (48) is polynomial as in (47), and in addition all matrix coefficients A_k have only integer entries: $A_k \in \operatorname{Mat}_{n\times n}(\mathbb{Z})$. Then, since the polynomial $q_0(t,x) = x_1$ also belongs to the subring $\mathbb{Z}[t,x]$, all subsequent polynomials q_k will also have integer coefficients, and their degrees will grow not faster than linearly in k. The growth of the norms $\|q_k\|$ can also be easily controlled.

Suppose that the length ℓ of the ascending chain of the corresponding ideals (55) is already known (in the univariate case it is relatively simple, see [**NY4**]). Then one can explicitly compute an upper bound r for the degrees of polynomial coefficients h_i in (56) in terms of ℓ, the degree $d = \deg A(t)$ and $n = \dim x$. To find the polynomials h_i themselves (or rather the univariate polynomials $a_i = h_i(\cdot, 0)$), it is now possible to use the method of indeterminate coefficients: writing each a_i as $\sum_{j=0}^{r} c_{ij} t^j$ and substituting them into the identity (57), we obtain a system of nonhomogeneous linear algebraic equations for the unknown variables $\{c_{ij} : i = 1, \dots, \ell, j = 0, \dots, r\}$, with integral coefficients matrix and the free terms column all explicitly bounded from below. For the same reason as in Example 4.3, in this "lattice" case explicit bounds for $\|a_i\|$ can be immediately produced in terms of n, d and r. ◀

The lattice polynomial system may look artificial, but in fact its appearance is natural. The explanation is given in the following principal example.

EXAMPLE 4.9. (universal system). All entries of the matrix coefficients A_k of the polynomial matrix function (47) can be considered as parameters and denoted by λ_i, $i = 1, \dots, (d+1)n^2$. With respect to these variables, the "universal matrix polynomial" $\sum_{k=0}^{d} A_k t^k$ is in fact in the "lattice" $\mathrm{Mat}_{n\times n}(\mathbb{Z}[t,\lambda])$, of known degree $d+1$: moreover, all matrix coefficients of $A(t)$ over the ring $\mathbb{Z}[t,\lambda]$ have only entries equal to zero or one.

In a similar way one can start iterations from the "general linear form" $q_0(t,x) = \sum_{i=1}^{n} \beta_i x_i$ which becomes a polynomial of degree $n+1$ with zero-one coefficients in $\mathbb{C}[x,\beta_1,\dots,\beta_n]$ and add the string of the coefficients $\{\beta_i\}$ to the parameter list. This would allow one to treat simultaneously isolated zeros of all nontrivial linear combinations. ◀

4.1.6. *Lengths of ascending chains.* Example 4.6 in fact proves that, for any dimension n and degree d of the matrix polynomial $A(t)$ from (47), one can majorize explicitly the number of isolated zeros of any linear combination $\sum \beta_i x_i(t)$ in terms of n, d, and $R = \sum_0^d \|A_k\|$ provided that the length of ascending chain of polynomial ideals (55) is explicitly known. We claim that this length is a "computable" function.

Notice that for each combination of n, d we have the single universal parameter space $\{\lambda\} = \mathbb{C}^{(d+1)n^2+n}$ (the coefficients of the matrix polynomial $A(t)$ and the initial linear form q_0), and the single chain of ideals in the polynomial ring $\mathbb{C}[t,x,\lambda]$ generated by polynomials q_k that in fact belong to the lattice $\mathbb{Z}[t,x,\lambda]$. Thus we have a well-defined function $\ell(n,d)$.

The construction of the generators q_k is absolutely explicit. The problem of verifying whether the next polynomial q_k belongs to the ideal I_{k-1} generated by the previous polynomials is constructive: there exists an algorithmic procedure allowing one to get a positive or negative answer in a finite number of steps for each k. Thus the length function ℓ is a "computable" (in the weakest sense) function of n and d: the above description can be transformed into an algorithm computing the value $\ell(n,d)$ for any given combination of n and d.

Coupled with the preceding discussion, this computability means that the problem on the number of zeros of functions defined by polynomial ODE's is algorithmically solvable. However, the complexity of this algorithm turns out to be very high in the sense that the bound on the growth rate of the function $\ell(n,d)$ is gargantuan and higher than any elementary function. In the next section we address this question in detail and show what other modifications are necessary in order to produce a "theoretically feasible" upper bound.

4.2. Lengths of chains.

4.2.1. *Descending chains of algebraic varieties.* The problem on ascending chains of polynomial ideals belongs entirely to the realm of commutative algebra. However, there is a parallel geometric problem that admits a considerably more transparent solution. Moreover, the construction used in the "geometric" proof can be adjusted (after introducing appropriate technical tools) to the algebraic case.

Consider a *strictly* decreasing chain of complex affine algebraic varieties

$$\mathbb{C}^n = X_0 \supset X_1 \supset X_2 \supset \cdots \supset X_k \supset \cdots, \tag{62}$$

where each variety X_k is given by a finite number of polynomial equations in $\mathbb{C}^n$ of degree no greater than d_k (no restrictions are imposed on the number or the structure of these equations, see [**H1**]). Without loss of generality we may assume that the sequence $d_1, d_2, \dots$ is nondecreasing.

By the same Noetherian property, the chain (62) must terminate after finitely many steps at some variety X_ℓ. The problem is to compute the length ℓ from the dimension n and knowing the bounds for the degrees d_k.

If the degrees d_k are all bounded, then the problem belongs to linear algebra, since the polynomial equations defining the varieties X_k will in fact constitute a finite-dimensional linear space, and its dimension will be the natural bound for the length of the chain (62). In order to avoid technical difficulties when talking about computable dependence on *infinite* input data $\{d_1, d_2, \dots, \}$, we will consider *finite-parameter* examples. The most important are the linear growth case, when

$$d_k = d + k, \quad k = 1, 2, \dots \tag{63}$$

or the exponential growth case

$$d_k = d^k, \quad k = 1, 2, \dots. \tag{64}$$

In both cases the natural number d is a parameter, and in both cases ℓ should be majorized in terms of n and d.

4.2.2. *Lexicographically decreasing sequences of words.* The advantage of the "geometric" problem on chains of complex algebraic varieties, is that it can be reduced to a purely combinatorial problem on lexicographically decreasing sequences of words.

Recall that any (affine algebraic) variety can be uniquely represented as the union of irreducible algebraic subvarieties of different dimensions varying from 0 (isolated points) to $n-1$ (hypersurfaces). Thus with any $X \subset \mathbb{C}^n$ one can associate a vector $\nu(X) \in \mathbb{Z}_+^n$ with n integer nonnegative coordinates $(\nu^{n-1}, \dots, \nu^1, \nu^0)$, where $\nu^i = \nu^i(X)$ stands for the *number* of irreducible i-dimensional components of X.

Denote by $\prec$ the lexicographic order on $\mathbb{Z}_+^n$, letting

$$(\nu^{n-1}, \dots, \nu^0) \prec (\bar\nu^{n-1}, \dots, \bar\nu^0)$$

if and only if for some $k = 1, \dots, n$

$$\nu^{n-1} = \bar\nu^{n-1}, \dots, \nu^k = \bar\nu^k, \quad \text{but } \nu^{k-1} < \bar\nu^{k-1}.$$

The following elementary observation is crucial.

LEMMA 10. *If $X \subsetneq Y \subset \mathbb{C}^n$, then $\nu(X) \prec \nu(Y)$.*

PROOF. This holds since:

(1) any irreducible component of X should belong to an irreducible component of Y, and
(2) if $A \subset B$ is a pair of *irreducible* varieties, then $\dim A \leqslant \dim B$ and in the case of equal dimensions necessarily $A = B$.

In other words, when passing from Y to X each irreducible component either completely survives or is split into a number of other components of strictly inferior dimensions. □

The fact that a descending chain of algebraic varieties must stabilize now follows from the following purely combinatorial claim.

PROPOSITION 5. *A lexicographically strictly decreasing chain*

$$\nu_1 \succ \nu_2 \succ \cdots \succ \nu_k \succ \cdots, \quad \nu_k \in \mathbb{Z}_+^n, \tag{65}$$

must be finite.

PROOF. For $n = 1$ the claim is obvious, since any decreasing sequence of non-negative integers must be finite. For an arbitrary n, the first "letters"[3] $\nu_k^{n-1} \in \mathbb{Z}_+$, $k = 1, 2, \dots$, form a nonincreasing sequence (it need *not* necessarily be strictly decreasing). However, no more than a finite number of values is taken. Along any interval of constancy of the first letter, the tails $(\nu_k^{n-2}, \dots, \nu_k^0)$ also form a lexicographically strictly decreasing sequence in $\mathbb{Z}_+^{n-1}$. Hence the length of each such segment is finite by the induction assumption, and the length of the whole chain is finite as the sum of finitely many finite numbers. □

This proof, being extremely simple, can be supplied by quantitative estimates of the lengths under additional assumptions on the *norms* of the words. Denote $\|\nu\| = \nu^{n-1} + \cdots + \nu^0$ the norm on $\mathbb{Z}_+^n$.

EXAMPLE 4.10. Assume that the norms of the words ν_k forming the chain (65) are bounded:

$$\|\nu_k\| \leqslant d + k, \quad \forall k = 1, 2, \dots \quad (d \text{ a natural parameter}). \tag{66}$$

Then the length of the chain as a function of n and d is bounded by a general recursive (but *not* primitive recursive) function. The explanation is as follows.

Let $f(n, d, i)$ be the maximal length of the chain under the additional constraint that the first "letter" of the first word is no greater than i, $\nu_1^{n-1} \leqslant i$. Then the restricted function $f(n+1, \cdot, i+1)$ can be expressed via $f(n, \cdot, \cdot)$ and $f(n+1, \cdot, i)$.

Indeed, the length of the *initial* segment of the chain, on which the first letter maintains its initial value i, can be at most $f(n, d, d)$, since the tails start with a word whose first letter can be at most d. After the first segment is exhausted, the remaining part of the chain begins with a word of length n, whose first letter is at most $i - 1$, and with the restriction on the norms of the words as follows,

$$\|\nu_k\| \leqslant d + \bigl(k + f(n, d, d)\bigr),$$

if the words of the remaining chain are numbered after the drop of the first letter. This is tantamount to replacing d by $f(n, d, d)$, therefore we obtain the recurrent inequality

$$f(n+1, d, i) \leqslant f(n, d, d) + f(n+1, d + f(n, d, d), i - 1). \tag{67}$$

Coupled with the boundary conditions

$$f(n, d, 0) \leqslant f(n-1, d, d), \quad f(1, d, i) \leqslant i,$$

this determines the upper bound for f completely. Notice the remarkable similarity between the recurrence rule (67) for f and that for the Ackermann generalized exponential (39). ◀

The growth rate determined by the above recursive function, is enormous: in terms of n, the dimension of the ambient space, the function f grows faster than any primitive function. On the other hand, this bound is essentially sharp: for lexicographically ordered chains it is rather obvious, for chains of polynomial ideals it was proved by G. Moreno [**MS2, MS1**], for chains of algebraic varieties the claim is apparently also true though not written down anywhere. In any case the bounds based on such an estimate of the length of chains should not be considered as constructive, even theoretically.

[3] We add the quotes saying "letters" rather than letters for the first time since the "words" ν_k are over the infinite "alphabet" $\mathbb{Z}_+$. However, the quotes will be dropped immediately.

However, under rather mild additional assumptions the bounds can be improved very considerably, in fact to become elementary (hence, in particular, primitive recursive) functions of n, d.

4.2.3. *Dynamically generated chains.* The chain of ideals (55) is generated by a dynamical system. More precisely, the rule (60) for the general case of a polynomial vector field (59) is a Lie derivative of the ring $\mathbb{C}[t, x]$, and the ideals I_k are generated by iterated Lie derivatives of the initial polynomial $q_0 \in \mathbb{C}[t, x]$.

A parallel construction for chains of algebraic varieties also involves a dynamical ingredient [**NY2, NY4**]. Namely, instead of the general chains (62) with the only restriction on the degrees of the varieties X_k, one should consider chains generated by a discrete time dynamical system in $\mathbb{C}^n$.

Let $F\colon \mathbb{C}^n \to \mathbb{C}^n$ be a polynomial map of some known degree d and $X \subset \mathbb{C}^n$ an algebraic subvariety. Consider the decreasing chain $\{X_k\}$ defined by the recursive rule

$$X_0 = X, \quad X_{k+1} = X \cap F^{-1}(X_k), \quad k = 0, 1, 2, \dots. \tag{68}$$

In other words,

$$X_k = X \cap F^{-1}(X) \cap F^{-2}(X) \cap \cdots \cap F^{-k}(X),$$

where F^k stands for the k times iterated map $F \circ \cdots \circ F$ and $F^{-k}(\cdot) = (F^k)^{-1}(\cdot)$ denotes the corresponding preimage.

Dynamically X_k can be described as the set of points $a \in \mathbb{C}^n$ such that the point a itself, together with its F-orbit $F(a), \dots, F^k(a)$ of length k, belong to X. This immediately implies that

$$F(X_k \setminus X_{k+1}) \subseteq X_{k-1} \setminus X_k, \quad k = 1, 2, \dots \tag{69}$$

(the differences occurring above consist of initial conditions of orbits that leave the variety X *exactly* after the specified number of steps).

The condition (69) means that the difference $X_k \setminus X_{k+1}$ cannot be too large compared with the preceding difference $X_{k-1} \setminus X_k$. For an arbitrary polynomial map this is not true, but under the additional assumption that F preserves dimensions of semialgebraic sets (i.e., takes curves into curves and not into points, though possibly creating singularities, and the same in higher dimensions), one can conclude that the dimensions of the differences $X_k \setminus X_{k+1}$ are nonincreasing:

$$\dim(X_k \setminus X_{k+1}) \leqslant \dim(X_{k-1} \setminus X_k), \quad k = 1, 2, \dots. \tag{70}$$

4.2.4. *Dynamically generated chains stabilize fast.* Consider a strictly descending chain of varieties (62) satisfying the additional condition (70) on the dimensions. Then the associated words $\nu_k = \nu(X_k) \in \mathbb{Z}^n_+$, in addition to the lexicographic decrease (65), display stronger monotonicity properties.

Consider the sequence of "heads" (as opposed to "tails"), obtained by truncating the words ν_k to their first s symbols, $[\nu_k]_s = (\nu_k^{n-1}, \dots, \nu_k^{n-s}) \in \mathbb{Z}^s_+$, for all $k = 1, 2, \dots$. For any lexicographically nonincreasing sequence, the sequence of heads of any fixed length s will again be nonincreasing (by definition of the lexicographic order). Yet this monotonicity can be nonstrict in general.

However, under the additional assumption (70) any truncated sequence of heads $\{[\nu_k]_s\}_{k=1}^{\infty}$ for the sequence $\nu_k = \nu(X_k)$ will be *strictly decreasing* in the following sense:

$$\forall s = 1, \dots, n-1 \quad [\nu_1]_s \succ [\nu_2]_s \succ \cdots \succ [\nu_{k-1}]_s = [\nu_k]_s = \dots \tag{71}$$

In other words, as soon as the first equality between truncated words occurs, the rest of the chain will have the same heads. Indeed, $[\nu_{k-1}]_s = [\nu_k]_s$ if and only if all irreducible components of X_{k-1} and X_k of dimensions $n - s$ and higher are the same, which means that the difference $X_{k-1} \setminus X_k$ is at most $(n-s-1)$-dimensional.

Such a type of descent ensures much faster convergence. Indeed, the first letter stabilizes after no more than $\|\nu_1\|$ steps, after which the problem is reduced to that for the tails, which are words of length $n - 1$. The corresponding inductive inequality is very simple.

EXAMPLE 4.11. Denote by $g(n, d)$ the maximal length of a decreasing sequence of words of length n growing no faster than linearly in the sense of the norm, $\|\nu_k\| \leqslant d+k$, with d being an integer parameter as in (63), under the additional assumption (71). Then the above observation implies that the length until stabilization of the first letter is at most d, whereas the sequence of tails of length $n - 1$ starts from the word of norm no greater than $d + d = 2d$ and hence its length by the induction assumption is no greater than $g(n - 1, 2d)$, so that finally

$$g(n, d) \leqslant d + g(n - 1, 2d), \quad g(1, d) \leqslant d. \tag{72}$$

This gives an upper bound for the length,

$$g(n, d) \leqslant (2^n - 1)d,$$

which is much better than in the general case considered in Example 4.10. ◀

EXAMPLE 4.12. The exponential bound (64) can be treated in a similar way. In this case the inductive inequality analogous to (72) takes the form

$$g(n, d) \leqslant d + g(n - 1, d^d), \tag{73}$$

which gives $g(n, d)$ as a tower function of height n. Though not elementary, this is obviously a primitive recursive function of both arguments. ◀

These simple examples illustrate the algorithmic complexity of the problem on lengths of lexicographically descending chains. For descending chains of algebraic varieties the answer follows in fact from Example 4.12, since the degrees of equations defining the algebraic varieties X_k grow exponentially: $\deg F^k = d^k$, where $d = \deg F$ (without loss of generality we may assume that $\deg X = d$ with the same d). The bounds for the norms $\|\nu(X_k)\|$ follow from a version of the Bézout theorem due to J. Heintz [**H3**].

The most technically difficult case is that of ascending chains of polynomial ideals, mainly since there is no uniqueness in the primary decomposition of such ideals, hence one cannot associate a word $\nu(I)$ with an ideal I, counting the number of primary components of various dimensions. However, the components of the maximal dimensions are correctly defined and may be counted, which allows one to implement a similar inductive proof. Details can be found in [**NY4**].

4.3. Restricted quasialgebraicity of Picard – Vessiot fields. Quasialgebraicity of a function field was defined as a property allowing for counting zeros globally. However, for technical reasons we need a weaker notion of *restricted quasialgebraicity*. Assuming $U \subset \mathbb{C}$ is a simply connected (usually polygonal or circular) domain containing no singular points on its boundary (but eventually some singularities *inside* U), we can restrict functions from $\mathbb{C}(X)$ on U. The result, denoted by $\mathbb{C}_U(X)$, is another function field, again consisting of multivalued functions.

DEFINITION 10. We say that $\mathbb{C}(X)$ is *restricted quasialgebraic in U* if the upper bounds on the number of isolated zeros in any triangle $T \subset U \setminus \Sigma$ can be given in the same terms as before, plus possibly some geometric parameters describing the relative position of Σ and U, most often the distance between Σ and the boundary ∂U.

4.3.1. *Bounds in the disk for polynomial systems.* The above discussion allows one to analyze completely the polynomial case.

THEOREM 8. *If the coefficient matrix $A(t)$ of a linear system* (31) *is polynomial as in* (47) *with bounded matrix coefficients of known total norm R, then the corresponding Picard – Vessiot extension $\mathbb{C}(X)$ will be quasialgebraic after restriction on any disk $D_r \subset \mathbb{C}$ of radius r.*

In other words, the bounds on the number of zeros of functions from $\mathbb{C}_{D_r}(X)$ will be explicit but depending on r. As $r \to +\infty$, the bounds blow up.

Blowing up of the bounds occurs for a very simple reason: the singular point at infinity is in general an irregular (and certainly non-Fuchsian) singularity which may be an accumulation point for isolated roots of polynomial combinations (see Section 3.1.4).

4.3.2. *Fuchsian system: reduction to the polynomial case.* To apply the results achieved for polynomial systems to the Fuchsian system (34), it is sufficient to introduce the new independent (complex time) variable.

The matrix function $A(t)$ from (34) can be reduced to the common denominator,

$$A(t) = \tfrac{1}{\chi(t)} P(t), \quad P = \sum_{i=0}^{d-1} P_i t^i, \quad \chi(t) = \prod_{i=1}^{d} (t - t_i).$$

The corresponding linear system is *orbitally* equivalent to the *polynomial* system

$$\begin{cases} \dot X = P(t)X, \\ \dot t = \chi(t), \end{cases} \qquad \cdot = \frac{d}{d\tau}, \tag{74}$$

where $\tau \in \mathbb{C}$ is the new time variable. The map

$$t \mapsto \tau(t) = \int_0^t \frac{dz}{\chi(z)}$$

is defined on the universal covering surface of $\mathbb{C} \setminus \Sigma$ and takes explicitly bounded values away from Σ. The inverse map $\tau \mapsto t(\tau)$ covers the complement $\mathbb{C} \setminus \Sigma$ so that for any triangle $T \subset \mathbb{C} \setminus \Sigma$ at distance $\varepsilon > 0$ from Σ one can find ρ such that T is covered by the image of the disk $\{|\tau| < \rho\} \subset \mathbb{C}$.

Application of Theorem 8 to the system (74) implies the following corollary (after some preliminary work).

COROLLARY 3. *The field $\mathbb{C}(X)$ constructed from the Fuchsian system* (34) *is quasialgebraic in any domain containing no singular points from Σ. The bounds would depend on $r = r(U) = \operatorname{dist}(\partial U, \Sigma) = \operatorname{dist}(\overline{U}, \Sigma)$ and blow up as $r \to 0^+$.*

4.3.3. *Rational systems with apparent singularities.* Actually, not all singular points are dangerous. As was already remarked, Theorem 7 is not about the number of zeros of solutions inside a triangle T, but rather about the variation of argument of these solutions along the boundary of T.

This means that the above theorems on restricted quasialgebraicity of Fuchsian systems away from their poles can be reformulated as assertions on computability of upper bounds for the variation of argument along polygonal paths distant from the singular locus Σ.

If the domain U has only *apparent* singularities inside, then all solutions of the system (31) are meromorphic in U and the order of their poles can be explicitly bounded in terms of the corresponding residue norms $\|A_j\|$. By the argument principle, this together with the bounds on the index along the boundary implies restricted quasialgebraicity of $\mathbb{C}_U(X)$.

This remark shows that it is the multivaluedness of solutions rather than any other circumstance that is an obstruction to global quasialgebraicity. In the next sections we show how multivalued functions can be treated, first locally and then globally.

5. Isomonodromic reduction principle and the Riemann–Hilbert problem

5.1. Isomonodromic fields: local theory.

5.1.1. *Euler field: an example.* The Euler field (36), obtained by adjoining to $\mathbb{C}$ all entries of the (multivalued) matrix function $t^A = \exp(A \ln t)$ solving the system (35), already occurred as a first example for which quasialgebraicity could be expected.

The necessary condition for quasialgebraicity is given in terms of the spectrum of the constant matrix A (the only residue of the system). We have already seen that if some of the eigenvalues are nonreal, then infinitely many real zeros may easily accumulate to the origin. It can be relatively easily seen that *under* this assumption on the spectrum of A, such accumulation is impossible not only for real, but also for complex zeros. Actually, a stronger assertion holds: the *number* of isolated roots can be explicitly majorized.

The following result (together with explicit bounds that in this case are rather accurate) can be found in [**KY**]. Consider a finite *set of exponents* $S \subset \mathbb{C}$ of diameter $\operatorname{diam} S = \max_{\lambda,\lambda' \in S} |\lambda - \lambda'|$ whose points may have nontrivial multiplicities $\nu(\lambda) \in \mathbb{N}$, so that $\sum_{\lambda \in S} \nu(\lambda) = \#S$.

THEOREM 9. *If $S \subset \mathbb{R}$, then the number of isolated roots of any finite sum*

$$\sum_{\lambda,k} c_{k\lambda}\, t^\lambda \ln^{k-1} t, \quad \lambda \in S,\ k \leqslant \nu(\lambda),\ c_{k\lambda} \in \mathbb{C}, \tag{75}$$

in any triangle $T \subset \mathbb{C} \setminus \{0\}$ is no greater than $\#S - 1 + 2 \operatorname{diam} S$. □

This theorem implies quasialgebraicity of the Euler field, since any element f from $\mathbb{C}[t^A]$ can be represented in the form (75) with explicit control over $\operatorname{diam} S$ and $\#S$ expressed in terms of $\|A\|$, $\dim A$ and $\deg f$.

The assumption of this theorem can be formulated in terms of the spectrum of the (only) monodromy operator.

COROLLARY 4. *If the spectrum of the monodromy operator M corresponding to a small loop around the origin belongs to the unit circle (i.e., all eigenvalues have modulus 1), then the Euler field $\mathbb{C}(t^A)$ is globally quasialgebraic, and the number of isolated zeros can be bounded in terms of the norm $\|A\|$.*

5.1.2. *Restricted quasialgebraicity near a Fuchsian point.* Consider a Fuchsian system (34) with a singular point $t_1 = 0$ at the origin and all residues of norm $\leqslant R$. Let ρ be a (sufficiently small) positive number such that all other singularities are at least 2ρ-distant and at most $1/\rho$-distant from the origin.

Let M be a monodromy matrix corresponding to a small positively oriented loop around the origin (this matrix is defined up to conjugacy) and $S \subset \mathbb{C}$ the spectrum of M (i.e., the collection of eigenvalues which is independent of anything but the system and the singular point).

THEOREM 10. *If $S \subset \{|\lambda| = 1\}$, then the field $\mathbb{C}(X)$ is quasialgebraic in the disk $D_\rho = \{|t| < \rho\}$. The bound on the number of isolated zeros can be given in terms of R and $1/\rho$.*

The outline of the proof of this theorem occupies the rest of this section. A similar result was proved in [**RY**] for *linear equations* of order n near a Fuchsian singular point.

5.1.3. *Joint fields.* Consider a Fuchsian singular point $t_1 = 0$ of the system (34) with the corresponding residue matrix $A_1 = B$. Then the fundamental matrix solution $X(t)$ in any disk D, containing no other singularities, can be represented as

$$X(t) = Y(t)\,t^B, \quad t \in D, \tag{76}$$

where Y is a meromorphic (single-valued) function of t in D. Replacing if necessary B by $B + rE$ with an appropriate $r \in \mathbb{Z}$, one can always consider $Y(t)$ as a holomorphic function at $t = 0$. Expanding all matrix products, we conclude that the field $\mathbb{C}(X)$ belongs to a bigger field $\mathbb{C}(Z, Y)$ spanned jointly by entries of the two matrix functions, $Z(t) = t^B$ and $Y(t)$. Whereas the field $\mathbb{C}(Z)$ is a Picard–Vessiot extension for the Euler system, this is not immediately obvious concerning the extension field $\mathbb{C}(Y)$. However, we claim that the field $\mathbb{C}(Y)$ is a subfield of a bigger Picard–Vessiot field for some other Fuchsian system having $t = 0$ as an apparent singularity. Indeed,

$$\dot{Y} = \dot{X}\,t^{-B} - t^{-1}Xt^{-B}\,B = A(t)Y - t^{-1}YB. \tag{77}$$

This is *not* a linear system for Y as a matrix function to satisfy (since it involves the matrix multiplication from both sides). But if all entries of Y are arranged as one column vector, then (77) becomes a system of n^2 linear ordinary differential equations with rational coefficients, exhibiting a Fuchsian singularity at $t = 0$ (actually, at all other poles of $A(t)$ as well). The residues of this larger-size system can be explicitly constructed and their norms bounded from above.

The joint field $\mathbb{C}(Z, Y)$ possesses a property that was already used for "single" extension fields.

LEMMA 11. *Variation of argument of any element from the joint field $\mathbb{C}(X, Y)$ along a polygonal path distant from the union of singular loci is a computable function. The same is true also for sufficiently small circular arcs around each singular point.*

Note that $t = 0$ by construction is an apparent singularity for this system. Thus the Picard–Vessiot field $\mathbb{C}(Y)$ is quasialgebraic restricted to D, as follows from Section 4.3.3.

The problem that we face now is to prove the *restricted* quasialgebraicity in D of the joint extension field $\mathbb{C}_D(Z,Y)$ containing $\mathbb{C}_D(X)$ by virtue of (76), having already established restricted quasialgebraicity of each of the fields $\mathbb{C}_D(Z)$ and $\mathbb{C}_D(Y)$ separately and the fact that Y has only one apparent singularity in D. This will be done by reducing the question on (restricted) quasialgebraicity of $\mathbb{C}_D(Z,Y)$ to that for $\mathbb{C}(Z)$, using the fact that Y has trivial monodromy in D. This reduction can be extracted from the papers by Petrov [**P2, P3**].

First we need some real analysis.

5.1.4. *Real closedness.* The real part of an analytic function f on an open domain U is not analytic (unless it is a constant). Yet the real part of the restriction $f|_\gamma$ on an analytic curve $\gamma \subset U$ can be extended to a neighborhood of γ and sometimes to the whole of U as an analytic function (of course, taking nonreal values outside of γ).

If $U = D$ is a disk centered at the origin and γ the real line $\mathbb{R}$, then for any function f meromorphic in D one can take

$$\operatorname{Re} f(t) = \tfrac{1}{2}\big(f(t) + \overline{f(\bar t)}\big), \quad \operatorname{Im} f(t) = \tfrac{1}{2i}\big(f(t) - \overline{f(\bar t)}\big), \tag{78}$$

and then $\operatorname{Re} f$ and $\operatorname{Im} f$ will again be meromorphic in D and equal to the real (resp. imaginary) part of f on $\mathbb{R}$.

This observation allows one to assume without loss of generality that all generators of the field $\mathbb{C}(Y)$ are real on the real axis. Otherwise, one should take their real and imaginary parts $\operatorname{Re} Y$ and $\operatorname{Im} Y$ as above (both are meromorphic matrix functions real on $\mathbb{R} \cap D$) and consider the field $\mathbb{C}(\operatorname{Re} Y, \operatorname{Im} Y)$ obviously containing $\mathbb{C}(Y)$. All properties of the field $\mathbb{C}(Y)$ are inherited by $\mathbb{C}(\operatorname{Re} Y, \operatorname{Im} Y)$.

For a multivalued function f ramified over the origin $t = 0 \in D$, taking its real or imaginary part on the whole real axis is an ambiguous operation because of the ramification: the real part on the positive semiaxis $\mathbb{R}_+$ can cease to be real after continuation on the negative semiaxis $\mathbb{R}_-$. Yet one can often guarantee that the field as a whole is closed by taking real/imaginary parts along any segment of the real axis.

LEMMA 12. *Let $A \in \operatorname{Mat}_{n\times n}(\mathbb{R})$ be a real constant matrix and $\mathbb{C}(Z)$, $Z(t) = t^A$, the corresponding Euler field. Then for any function $f \in \mathbb{C}(Z)$ its real or imaginary part can be extended from the positive semiaxis $\mathbb{R}_+$ or the negative semiaxis $\mathbb{R}_-$ to functions $g_\pm$ again belonging to $\mathbb{C}(Z)$:*

$$\forall f \in \mathbb{C}(Z)\ \exists g_\pm \in \mathbb{C}(Z)\colon \quad \operatorname{Re} f|_{\mathbb{R}_\pm} = g_\pm|_{\mathbb{R}_\pm},$$

and the same for the imaginary parts $\operatorname{Im} f|_{\mathbb{R}_\pm}$.

We will denote the functions $g_\pm = \operatorname{Re} f|_{\mathbb{R}_\pm}$ and their imaginary counterparts extending $\operatorname{Im} f|_{\mathbb{R}_\pm}$ by $\operatorname{Re}_\pm f$ and $\operatorname{Im}_\pm f$ respectively.

PROOF. The field $\mathbb{C}(Z)$ is independent of the choice of the fundamental solution, hence for each semiaxis $\mathbb{R}_\pm$ one can choose a solution $Z_\pm$ of the Euler system $t\dot Z = AZ$ that is real on that semiaxis (recall that the matrix A is real so $A(t) = t^{-1}A$ is real-valued on $\mathbb{R}$). Then it remains only to define $g_\pm$ using the identities

$$\operatorname{Re}\Big(\sum c_\alpha Z_\pm^\alpha\Big) = \sum_\alpha (\operatorname{Re} c_\alpha) Z_\pm^\alpha \in \mathbb{C}[Z], \quad c_\alpha \in \mathbb{C},$$

and similarly for the imaginary part (with obvious modifications for the field of fractions $\mathbb{C}(Z)$). □

REMARK. The fact that the matrix A is real is not a restriction: one can always consider a bigger Euler field of dimension $2n$ with the real block-diagonal matrix $\operatorname{diag}(\operatorname{Re} A, \operatorname{Im} A)$.

5.1.5. *Variation of argument and zeros of imaginary part.* The following elementary statement will play the central role in the constructions below.

LEMMA 13. *If $f\colon \mathbb{R} \supset [a,b] \to \mathbb{C}$ is a complex-valued function having no zeros on the real interval $[a,b]$, then variation of argument of f along this interval is no greater than $\pi(\#\{\operatorname{Im} f = 0\} + 1)$.*

Note that here no extension from $[a,b]$ is required, hence $\operatorname{Im} f$ stands for the usual imaginary part of the restriction.

PROOF. If $|\operatorname{Arg} f(t_1) - \operatorname{Arg} f(t_2)| > \pi$, then the imaginary part $\operatorname{Im} f$ must vanish somewhere between t_1 and t_2 on $[a,b]$. □

Actually, this result is true for any linear combination $\alpha \operatorname{Re} f + \beta \operatorname{Im} f$. The function f may be allowed to have isolated zeros (but certainly not vanish identically), provided that zeros of $\operatorname{Im} f$ are counted with multiplicities.

5.1.6. *Isomonodromic reduction: the local case.* Now everything is ready to prove that quasialgebraicity of $\mathbb{C}(Z)$ implies the restricted quasialgebraicity of $\mathbb{C}(Z,Y)$ in $D = \{|t| < 1\}$, provided that $Y(t)$ is single-valued in D and satisfies a Fuchsian system with bounded residues and all singularities ε-distant from ∂D.

Any function $f \in \mathbb{C}(Z,Y)$ can be written in D as

$$f(t) = \sum_1^N z_i(t) y_i(t), \quad z_i \in \mathbb{C}(Z),\ y_i \in \mathbb{C}_D(Y),\ i = 1, \dots, N, \tag{79}$$

where y_i are *monomials* from the field $\mathbb{C}(Y)$ and hence real on $\mathbb{R}$, and the number of terms N as well as the degrees $\deg z_i, \deg y_i$ (with respect to the corresponding fields $\mathbb{C}(Z)$ and $\mathbb{C}_D(Y)$) are explicitly computable in terms of $\deg f$ with respect to $\mathbb{C}(X) \subset \mathbb{C}(Z,Y)$. Indeed, for each generator X_{ij} of degree 1 in $\mathbb{C}(X)$ such a representation can be chosen with no more than n terms, each being bilinear in Y and Z.

Assume that the field $\mathbb{C}(Z)$ is quasialgebraic. Then division by z_1 changes the number of zeros by no more than some known number. Hence when counting the number of zeros of f in any triangle $T \subset D \setminus \{0\}$, one can assume that $z_1(t) \equiv 1$, that is, $f = y_1 + \sum_2^N z_i y_i$.

Also, without loss of generality we may assume that T does not intersect the real axis (this can be also always achieved by subdividing it into smaller triangles if necessary), and belongs to the upper half-plane. Then the number of zeros of f in T is no greater than the sum of four terms, variation of argument of f along the sufficiently small arc $\{|t| = \varepsilon_0 \ll 1,\ \operatorname{Im} t > 0\}$ around the origin, and similar contributions from the semicircle $\{|t| = 1,\ \operatorname{Im} t > 0\}$, and two rectilinear intervals $[-1, -\varepsilon_0) \subset \mathbb{R}_-$ and $(\varepsilon_0, 1] \subset \mathbb{R}_+$ respectively.

The contributions from the two circular arcs are computable functions by Lemma 11. As for the two rectilinear segments, the contribution of each of them is bounded by the number of zeros of $\operatorname{Im} f$ on them. By the assumption on the matrix function Y, $\operatorname{Im} y_i \equiv 0$, hence

$$\operatorname{Im}_{\pm} f = \operatorname{Im} y_1 + \sum_{2}^{N} y_i \operatorname{Im} z_i = \sum_{2}^{N} y_i \operatorname{Im}_{\pm} z_i.$$

By Lemma 12, the field $\mathbb{C}(Z)$ is closed under taking imaginary parts, hence for all i the functions $\operatorname{Im}_{\pm} z_i$ are again in $\mathbb{C}(Z)$. Thus the question on the number of zeros of f represented by (79) in T is reduced to that for the number of zeros of two functions $\operatorname{Im}_{+} f, \operatorname{Im}_{-} f \in \mathbb{C}(Z, Y)$ (one for each semiaxis), each of them involving fewer terms (at most $N-1$). This allows one to continue the process inductively, reducing the problem for zeros of f to that for some 2^N functions from $\mathbb{C}(Z)$. In the last step the number of isolated zeros of a product $z_N y_N$ is majorized by the sum of zeros of each term (known since quasialgebraicity of $\mathbb{C}(Z)$ and $\mathbb{C}(Y)$ is already established).

REMARK. The above described construction reducing quasialgebraicity of the joint field $\mathbb{C}_D(Z, Y)$ to that of $\mathbb{C}_D(X)$ if Y has trivial monodromy in D, is very similar to the standard differentiation-division scheme (based on the Rolle lemma) used to obtain bounds on the number of isolated zeros of real functions. Here the role of the differentiation is played by the operators $\operatorname{Im}_{\pm}$, whereas single-valued functions real on $\mathbb{R}$ play the role of constants killed by differentiation.

5.1.7. *Nonuniform quasialgebraicity.* From Theorem 10 one can immediately derive nonuniform quasialgebraicity of Fuchsian systems. Consider such a system with d singular points.

Notice that the residue matrices A_j are invariant under conformal automorphisms of the independent variable, hence the residual norm of the corresponding matrix function $R = \max_{i=1,\dots,d} \|A_i\|$ is also invariant.

Using such conformal automorphisms, one can always place any three poles of $A(t)$ at any three points, say, $0, 1$ and ∞, but starting from the fourth pole one has a nontrivial parameter characterizing the spread of singular points on the sphere. Let ρ be a small positive number such that:

(1) $|t_i - t_j| \geqslant 2\rho$, $i, j = 1, \dots, d$, $i \neq j$;
(2) $|t_i| \leqslant 1/\rho$, $i = 1, \dots, d$.

Consider the monodromy matrices M_j corresponding to small loops going around t_j. As before, their spectra S_j are uniquely defined.

THEOREM 11. *If all spectra S_j belong to the unit circle, then the field $\mathbb{C}(X)$ is quasialgebraic. The bound for the number of zeros can be given in terms of R and ρ.*

PROOF. Draw disjoint circles of radius ρ around each singularity and of radius $1/\rho$ around the origin (this circle bounds a neighborhood of infinity on $\mathbb{C}P^1$). Restricted to each circle, the field is quasialgebraic by Theorem 10. On the complement there are no singularities, so after triangulation of this multiply connected domain one can apply Corollary 3. □

5.2. Uniform quasialgebraicity of Fuchsian systems. In this section we briefly explain additional work to be done in order to obtain the bounds for quasialgebraicity that would be independent of the relative position of singular points. In other words, we look for bounds that would remain explicit and uniform over ρ as the latter tends to zero.

5.2.1. *Isomonodromic reduction: the general case.* The constructions of the previous section can be easily modified for a more general situation.

The *isomonodromic reduction principle* asserts that the property of a Picard–Vessiot extension $\mathbb{C}(X)$ to be quasialgebraic in a domain $U \subset \mathbb{C}$ (in the sense of Definition 7) depends only on the monodromy of the corresponding system along the loops entirely belonging to U and not on what happens outside U.

More precisely, two Fuchsian systems with fundamental solutions $X(t)$ and $Z(t)$ are called *isomonodromic relative to a domain* U if they have the same singular locus Σ in U and monodromy matrices M_γ for all loops entirely belonging to U, are the same for the two fundamental solutions.

Assume that the residues of the two coefficient matrices, $A(t) = \dot{X}(t)X^{-1}(t)$ and $B(t) = \dot{Z}(t)Z^{-1}(t)$ (all of them, including those at the singular points outside U) are all explicitly bounded. Assume also that U is a polygonal domain (say, a triangle) and the boundary ∂U is away from all singularities.

Then the isomonodromic reduction principle asserts that the two fields $\mathbb{C}(X)$ and $\mathbb{C}(Z)$ after restriction to U are both quasialgebraic or not quasialgebraic simultaneously.

The proof in the case when all singularities in U fall on one straight line is very similar to the local case. Namely, consider the matrix fraction $Y(t) = X(t)Z^{-1}(t)$ possessing trivial monodromy in U and embed $\mathbb{C}(X)$ into the joint field $\mathbb{C}(Z, Y)$. Writing elements of this joint field as $\sum y_i z_i$ and applying the above algorithm of alternating division and taking the imaginary parts, the question on the number of zeros of $f \in \mathbb{C}(X)$ can be reduced to that for several auxiliary functions from $\mathbb{C}(Z)$.

5.2.2. *Inductive strategy.* The isomonodromic reduction principle as described above would allow for an inductive proof of the uniform quasialgebraicity of Fuchsian systems if one could always construct a Fuchsian system that would be isomonodromic to a given one in a specified simply connected domain, while having no other singularities outside this domain.

The inductive proof may look as follows. For Fuchsian system with only two singularities (Euler systems) the quasialgebraicity is known. Assume that it is already established for all systems with fewer than d finite singularities, and consider a system with d finite singular points forming the locus $\Sigma \subset \mathbb{C}$.

As was already noticed, it is impossible to make a conformal transformation placing all d points of Σ well apart from each other. However, one can always achieve a situation in which all finite singular points form a set of diameter *exactly* 1 inside the disk of radius 1 centered at the origin. In this case one can draw a line that is at least $1/2d$-distant from all points of Σ and such that to each side of this line lies at least one (hence at most $d-1$) point(s) of Σ.

Now one can easily construct two polygonal domains each containing no more than $d-1$ points of Σ, together covering the whole of Σ and with boundaries distant from both Σ and infinity.

Assume that for each such domain U a Fuchsian system can be found so that it will be isomonodromic with the given one in U, while still having the residual norm bounded in terms of the residual norm of the initial system.

Then application of the isomonodromic reduction principle would allow one to reduce the question on quasialgebraicity of the initial system in U to that for a

Fuchsian system with $\leqslant d-1$ finite singular points. By the inductive assumption, the latter question can be explicitly answered.

The only assumption to monitor along this inductive process is that on eigenvalues of the monodromy operators. Clearly, one should assume that the spectral condition (on unit absolute values of eigenvalues) holds for all small loops around singular points. However, it is not sufficient, since this condition does not survive the above surgery (keeping the singularities inside U and replacing the rest by a single regular point). Indeed, after replacing all singularities outside U by one singular point at infinity, we create a point whose local monodromy coincides with that of the boundary ∂U. Thus one must additionally assume that the monodromy along the boundary of U must also satisfy the spectral condition.

Unfortunately, there is no way to predict how the partition into distant "clusters" of singular points will proceed when carried out inductively. Instead it is sufficient to assume that the spectral condition is satisfied for any simple loops (geometrically non-selfintersecting closed Jordan curves).

This conditional construction would prove the following theorem.

THEOREM 12 (see [**NY6**]). *If the monodromy operators along all simple loops have only eigenvalues of modulus* 1, *then the Fuchsian system is uniformly quasialgebraic. Upper bounds for the number of isolated zeros can be given in terms of the number of singular points and the residual norm of the system, uniformly over all configurations of the singularities.*

However, the arguments above give only a general idea of how the actual proof in [**NY6**] is organized. The difficulties are of two kinds, technical and fundamental.

An example of the technical problem is the isomonodromic reduction principle: it was formulated (and is actually proved) for a particular case when all singular points lie on just one straight line (which can then be identified with the real axis). However, it is sufficient for the purposes of the proof after suitable preparation of the initial Fuchsian system.

The fundamental problem concerns the (im)possibility of constructing a Fuchsian system with the prescribed monodromy. It was recently discovered that there exist obstructions to solvability of this problem, which have to be somehow circumvented. In addition, one has to readdress the proof of the corresponding positive results so that they would become *constructive*, yielding bounds for the residual norm of the constructed systems.

5.3. Quantitative Riemann – Hilbert problem of matrix factorization.

5.3.1. *Riemann – Hilbert problem: background.* Given d distinct points $t_1, \dots, t_d$ on the Riemann sphere $\mathbb{C}P^1$ and d invertible matrices $M_1, \dots, M_d$ satisfying the identity $M_1 \dots M_d = E$ (the identity matrix), construct a Fuchsian system having singular points at $t_1, \dots, t_d$ and only there, for which the matrices M_j would be monodromy factors for some fundamental solution.

This is the strongest form of the problem known as Hilbert's 21st problem or the *Riemann – Hilbert problem.* Here are some of many known results concerning its solvability.

Plemelj theorem. For any collection of points and any monodromy matrices, one can construct a linear system with all but one singularity Fuchsian; the last singular point is regular and can be made Fuchsian if the corresponding monodromy matrix is diagonalizable. As a corollary, one can always construct a Fuchsian system with one extra singular point that would be an apparent singularity [**P4, AI**].

Bolibrukh – Kostov theorem. If the monodromy group generated by the matrices M_j is irreducible, then the Riemann – Hilbert problem is solvable [**B3, K**].

Bolibrukh counterexamples. There exists a reducible group generated by three 4×4-matrices which cannot be realized as a monodromy group of a Fuchsian system [**B2**].

What is missing in all these formulations, is the possibility of placing any upper bound on the norms of the residue matrices of the corresponding Fuchsian system (provided that the problem is solvable).

5.3.2. *Surgery.* In this section we formulate an analytic problem that is sufficient to solve in order to construct a system isomonodromic to the given one in a domain $U \subset \mathbb{C}$ with all singularities away from the boundary.

To construct such a system, we need to remove all finite singularities of the initial system from $\mathbb{C} \backslash U$, leaving only one singular point at infinity, without affecting the monodromy of the system in U.

For simplicity we assume that U is the unit disk centered at the origin, and the annulus $K = \{\frac{1}{2} < |t| < 2\}$ is free of singular points (i.e., $\Sigma \cap K = \varnothing$). Denote by $X(t)$ the (multivalued) fundamental matrix solution. In general, X is multivalued even in K, however, for an appropriate constant matrix B the product $W(t) = Xt^{-B}$ is a single-valued matrix function that is holomorphic and holomorphically invertible in K.

Suppose that the matrix function $W = X(t)\,t^{-B}$, holomorphic and invertible in K, can be represented as the matrix ratio of two other matrix functions, $H_0(t)$ and $H_\infty(t)$, so that:

(1) H_0 is holomorphic and holomorphically invertible in the disk $D_0 = \{|t| < 2\}$;
(2) $H_\infty(t)$ is holomorphic and holomorphically invertible in the disk $D_\infty = \{|t| > \frac{1}{2}\}$, including the point $t = \infty$;
(3) on the intersection $K = D_0 \cap D_\infty$,

$$X(t)\,t^{-B} = H_0^{-1}(t) H_\infty(t), \quad t \in D_0 \cap D_\infty. \tag{80}$$

Then the two expressions, $X_0(t) = H_0(t)X(t)$ defined in D_0 and $X_\infty(t) = H_\infty(t)\,t^B$ defined on D_∞, agree on the intersection and hence their "logarithmic derivatives," $B_0(t) = \dot{X}_0(t)X_0^{-1}(t)$ and $B_\infty(t) = \dot{X}_\infty(t)X_\infty^{-1}(t)$, coincide on K and together define a single-valued meromorphic function $B(t)$ on the entire sphere $\mathbb{C}P^1$.

The poles of $B(t)$ can be easily described: due to the holomorphic invertibility of $H_{0,\infty}(t)$ they can occur either at the poles of $A(t)$ that are inside D_0, or at $t = \infty$. In both cases the residues can be easily computed: for any $t_j \in \Sigma \cap D_0$,

$$B(t) = H_0(t_j) \cdot \frac{A_j}{t - t_j} \cdot H_0^{-1}(t_j) + \cdots,$$

where the dots stand for terms holomorphic at t_j. In a similar way, at infinity we have

$$B(t) = H_\infty(\infty) \cdot \frac{B}{t} \cdot H_\infty^{-1}(\infty) + O(1/t^2),$$

where $O(1/t^2)$ stands for a holomorphic matrix function of the corresponding growth. This means that in fact

$$B(t) = \sum_{t_j \in D_0} \frac{B_j}{t - t_j}, \qquad B_j = H_0(t_j)\, A_j\, H_0^{-1}(t_j), \tag{81}$$

and $\sum_j B_j = H_\infty(\infty)\, B\, H_\infty^{-1}(\infty)$.

The Fuchsian system with the matrix of coefficients $B(t)$ given by (81) would serve our purposes, since its solution $H_0 X$ has the same monodromy in D_0 as the initial matrix solution X.

In order to complete the proof, one would have to obtain upper bounds on the matrices $H_j = H_0(t_j)$ and their inverses H_j^{-1} for all singular points $t_j \in \Sigma \cap D_0$. This would imply an upper bound on the residual norm of the matrix function $B(t)$ and finally would allow for the inductive proof as explained in Section 5.2.2.

5.3.3. *Birkhoff - Grothendieck factorization.* Unfortunately, finding factorization (80) satisfying all properties above is impossible.

EXAMPLE 5.1. One obstruction to holomorphic factorization can be immediately seen. Consider the determinant $w(t) = \det W(t)$: this is a holomorphic invertible function in the annulus K, and variation of the argument of this function along, say, the middle circle of the annulus is an integer ν. If this number is nonzero, then either $\det H_0$ or $\det H_\infty$ must have zeros and/or poles in the respective domains D_0, resp. D_∞.

If $n = 1$, i.e., all matrices are of size 1×1, then one can always achieve factorization of the form

$$W(t) = H_0^{-1}(t)\, t^\nu\, H_\infty(t) \tag{82}$$

with 1×1-matrices H_0, H_∞ holomorphically invertible in the respective domains. ◀

Both the positive and the negative assertions present in the above example admit generalization for the general n-dimensional matrix case.

THEOREM 13 (see [**GK**]). *A matrix function $W(t)$ holomorphic and invertible in the annulus K can be factorized as follows,*

$$W(t) = H_0^{-1}\, t^G\, H_\infty(t), \quad G = \operatorname{diag}(\nu_1, \dots, \nu_n), \nu_i \in \mathbb{Z}, \tag{83}$$

with the matrix factors H_0, H_∞ holomorphic and invertible in D_0, resp. D_∞.

The integers ν_i, called partial indices, are uniquely determined (the same for all representations with the above properties).

The decomposition (83) can be used for surgery if we incorporate the term t^G into $H_\infty(t)$. Then the matrix function $Y(t)$ defined as $H_0 X$ in D_0 and as $t^G H_\infty\, t^B$ in D_∞ satisfies a linear system of ordinary differential equations with only Fuchsian singular points in D_0 and a regular non-Fuchsian point $t = \infty$. The principal Laurent part of the matrix $B(t) = \dot Y Y^{-1}$ at $y = \infty$ can be easily described: the multiplicity of the pole at infinity is bounded in terms of $\nu = \|G\| = \max_j |\nu_j|$, and the magnitude (norm) of the Laurent coefficients of B at $t = \infty$ is bounded in terms of the Taylor coefficients of order $\leqslant \nu$ of the matrix H_∞ and its inverse.

Actually, the form of the matrix $t^G H_\infty(t)$ is not important: in order to have a system with a regular singularity at infinity, it is sufficient to have a matrix factorization

$$W(t) = H_0^{-1} H_\infty, \quad H_\infty(t)^{\pm 1} = C_{\pm,0}(t) + \sum_{j=1}^{\nu} C_{\pm,j} t^j, \tag{84}$$

with holomorphic invertible matrix H_0 and meromorphic invertible matrix function $H_\infty(t)$ having (together with its inverse H_∞^{-1}) the only pole at $t = \infty$ of order $\le \nu$ with the Laurent (matrix) coefficients $C_{\pm,j}$.

REMARK. The assertion of Theorem 13 is certainly not the strongest known. The matrix function $W(t)$ can be defined only on the middle circle $\{|t| = 1\}$ of the annulus K and be rather weakly regular on it, and still the factorization will be possible then, with the terms H_0, H_∞ holomorphic invertible inside (resp., outside) the circle. Some noncircular contours can be also allowed.

5.3.4. *Modified surgery.* One can easily modify the structure of the inductive construction above to cover the case of systems having only Fuchsian finite singular points and a regular possibly non-Fuchsian singularity at $t = \infty$. Such a system has matrix of coefficients that can always be written as

$$A(t) = \sum_{j=1}^{d} \frac{A_j}{t - t_j} + \sum_{i=0}^{\nu} B_i t^i, \tag{85}$$

and the residual norm for such systems should be defined as

$$\|A(\cdot)\| = \sum_j \|A_j\| + \sum_i \|B_i\|. \tag{86}$$

The surgery described in Section 5.3.2, using the factorization (84), allows one to pass from one system from such a class to another system from the same class, having no finite singularities outside D_0. In order to carry out inductively the bounds for zeros, one has to majorize the magnitude of all Laurent coefficients of the new system in terms of the norm $R = \|A(\cdot)\|$ of the initial system.

5.3.5. *Bounds.* Suppose that a system from the class (85) of explicitly bounded norm $R = \|A(\cdot)\|$ has no singularities in the annulus $K = \{\frac{1}{2} < |t| < 2\}$. Then the following bounds can be explicitly computed.

(1) the norm of the monodromy M corresponding to the circle $\{|t| = 1\} \subset K$ and its matrix logarithm B;
(2) the pointwise upper bound for norm of the fundamental solution $X(t)$ with $X(1) = E$ and its inverse $X^{-1}(t)$ in any smaller annulus K', say, $\{\frac{2}{3} < |t| < \frac{3}{2}\}$;
(3) the pointwise upper bound for $\|W(t)\| + \|W^{-1}(t)\|$ in the smaller annulus K'.

In order to estimate the Laurent coefficients at all singular points after the surgery, it would be sufficient to find the factorization (84) and supply the following bounds,

(1) $\max_{t \in D_0} \|H_0(t)\| + \|H_0^{-1}(t)\|$ (this would allow one to estimate the norms of residues at all finite singularities),
(2) the bound ν for the order of the pole of $H_\infty^{\pm 1}$;
(3) $\max_{t \in D_\infty} \|C_{+,0}(t)\| + \|C_{-,0}(t)\|$ together with
(4) $\sum_{i=1}^{\nu} \|C_{\pm,i}\|$ to majorize the norms of all Laurent coefficients at infinity.

5.3.6. *Quantitative matrix factorization.* Unfortunately for our purposes, the known methods of constructing the Birkhoff–Grothendieck factorization (83), based on index theory for integral operators, do not allow for quantitative conclusions. Moreover, in some sense the problem admits no solution. The reason for this is the known *instability* of the partial indices $\nu_1, \dots, \nu_n$: an arbitrarily small variation (in the uniform norm) of the function W can result in a jump of the partial indices. This is clearly incompatible with the existence of any bounds that would be continuous in the C^0-norm.

To regain control over the "magnitude" of the factors, one has to renege on the specific form of the Birkhoff–Grothendieck factorization and require only that the "exterior" factor $H_\infty(t)$ together with its inverse $H_\infty^{-1}(t)$ both have a pole at $t = \infty$ with the order and all Laurent coefficients explicitly bounded from above.

THEOREM 14 (Novikov and Yakovenko [**NY6**]). *An $n \times n$-matrix function $W(t)$ holomorphic and holomorphically invertible in the annulus*

$$K = \{(1+2\varepsilon)^{-1} < |t| < (1+2\varepsilon)\}, \quad \varepsilon > 0, \tag{87}$$

and bounded together with its inverse there,

$$\|W(t)\| + \|W^{-1}(t)\| < R < +\infty, \quad t \in K, \tag{88}$$

can be factorized as $W(t) = H_0^{-1} H_\infty(t)$ with the matrix functions H_0, H_∞ satisfying the following conditions.

1. *$H_0(t)$ is holomorphic invertible in the disk $D_0' = \{|t| < (1+\varepsilon)\}$ and satisfies the inequality $\|H_0(t)\| + \|H_0^{-1}(t)\| \leqslant R'$ in this disk,*
2. *$H_\infty(t)$ is holomorphic and holomorphically invertible in the complement $D_\infty' = \{(1+\varepsilon)^{-1} < |t| < +\infty\}$ and both H_∞ and H_∞^{-1} have at most a pole of order ν at $t = \infty$;*
3. *the coefficients $C_{\pm,i}$ of the Laurent expansions*

$$P_\pm(t) = H_\infty^{\pm 1}(t) = \sum_{i=0}^{\nu} C_{\pm,i} t^i$$

are all bounded by R' in the sense of the matrix norm;

4. *the "regular parts" are bounded uniformly in D_∞' so that $\|H_\infty^+ - P_+(t)\| + \|H_\infty^- - P_-(t)\| \leqslant R'$ there.*

The integer parameter ν and real parameter R' can be expressed as explicit elementary functions of n, R and ε, the "width" of the annulus K.

5.3.7. *Conclusion.* Theorem 14 provides the last tool necessary to run the inductive proof and construct an explicit primitive recursive upper bound for the number of isolated roots of solutions to Fuchsian systems. The bounds can be explicitly written down in the sense that all primitive recursions describing them can be extracted from the constructions above. However, this has not been done, among other reasons, because of the very excessive bounds that appear in this way. Besides nested inductive constructions, this occurs because of very excessive bounds on the lengths of ascending chains of polynomial ideals (see Section 4.1). However, some very recent works of A. Grigoriev suggest that this crucial step can be considerably improved and, instead of a tower of four stories, something like a double exponential bound can be achieved, at least for linear systems with rational coefficients. This improvement may affect all other parts of the global construction as well, resulting in a considerably more realistic quasialgebraicity-type statements.

Acknowledgements. The main theme of these lecture notes was surveying several recent results obtained jointly with Dmitry Novikov, who discussed with me the text on numerous occasions and greatly contributed to putting it in final form.

I am grateful to Dana Schlomiuk who invited me to deliver these lectures on the Workshop organized in Université de Montreal in June – July 2000. Part of these notes was written during my sabbatical stay in University of Toronto. During these periods I tried various ways of exposing parts of these notes and collected remarks and advises from many listeners, among them A. Bolibrukh, Yu. Ilyashenko, M. Jakobson, V. Katsnelson, A. Khovanskiĭ, V. Matsaev, Ch. Miller, P. Milman, B. Schapiro, P. Speisegger, Y. Yomdin.

Finally, I am very much indebted to the colleagues who read preliminary versions of these notes and pointed to numerous errors, especially to Christiane Rousseau and Iliya Iliev. Of course, the responsibility for the remaining bugs remains entirely with the author.

References

[AGZV] V. I. Arnol′d, S. M. Guseĭn-Zade, and A. N. Varchenko, *Singularities of differentiable maps. II: Monodromy and asymptotics of integrals*, Monogr. Math., vol. 83, Birkhäuser, Boston, MA, 1988.

[AI] V. I. Arnol′d and Yu. S. Ilyashenko, *Ordinary differential equations*, Current Problems in Mathematics. Fundamental Directions, Vol. 1, Itogi Nauki i Tekhniki, Akad. Nauk SSSR, Vsesoyuz. Inst. Nauchn. i Tekhn. Inform., Moscow, 1985, pp. 7 – 149 (Russian); English transl., Dynamical Systems. I, Encyclopaedia Math. Sci., vol. 1, Springer, Berlin, 1988, pp. 1 – 148.

[B1] R. I. Bogdanov, *Bifurcations of a limit cycle of a certain family of vector fields on the plane*, Trudy Sem. Petrovsk. **2** (1976), 23 – 35 (Russian).

[B2] A. A. Bolibrukh, *The Riemann – Hilbert problem*, Uspekhi Mat. Nauk **45** (1990), no. 2(272), 3 – 47 (Russian); English transl., Russian Math. Surveys **45** (1990), no. 2, 1 – 58.

[B3] ______, *Sufficient conditions for the positive solvability of the Riemann – Hilbert problem*, Mat. Zametki **51** (1992), no. 2, 9 – 19 (Russian); English transl., Math. Notes **51** (1992), no. 1-2, 110 – 117.

[B4] ______, *The Riemann – Hilbert problem and Fuchsian differential equations on the Riemann sphere*, Proceedings of the International Congress of Mathematicians, Vol. 1, 2 (Zürich, 1994), Birkhäuser, Basel, 1995, pp. 1159 – 1168.

[BD] P. Bonnet and A. Dimca, *Relative differential forms and complex polynomials*, Bull. Sci. Math. **124** (2000), no. 7, 557 – 571.

[BFY1] M. Briskin, J.-P. Françoise, and Y. Yomdin, *The Bautin ideal of the Abel equation*, Nonlinearity **11** (1998), no. 3, 431 – 443.

[BFY2] ______, *Une approche au problème du centre-foyer de Poincaré*, C. R. Acad. Sci. Paris Sér. I Math. **326** (1998), no. 11, 1295 – 1298.

[BFY3] ______, *Center conditions, compositions of polynomials and moments on algebraic curves*, Ergodic Theory Dynam. Systems **19** (1999), no. 5, 1201 – 1220.

[DRR] F. Dumortier, R. Roussarie, and C. Rousseau, *Hilbert's* 16*th problem for quadratic vector fields*, J. Differential Equations **110** (1994), no. 1, 86 – 133.

[É] J. Écalle, *Introduction aux fonctions analysables et preuve constructive de la conjecture de Dulac*, Actualités Math., Hermann, Paris, 1992.

[F1] O. Forster, *Lectures on Riemann surfaces*, Grad. Texts in Math., vol. 81, Springer-Verlag, New York, 1991.

[F2] J.-P. Françoise, *Successive derivatives of a first return map, application to the study of quadratic vector fields*, Ergodic Theory Dynam. Systems **16** (1996), no. 1, 87 – 96.

[FP] J.-P. Françoise and C. C. Pugh, *Keeping track of limit cycles*, J. Differential Equations **65** (1986), no. 2, 139 – 157.

[G1] A. Gabrielov, *Multiplicity of a zero of an analytic function on a trajectory of a vector field*, The Arnoldfest (Toronto, 1997), Fields Inst. Commun., vol. 24, Amer. Math. Soc., Providence, RI, 1999, pp. 191–200.

[G2] L. Gavrilov, *Petrov modules and zeros of Abelian integrals*, Bull. Sci. Math. **122** (1998), no. 8, 571–584.

[G3] ———, *The infinitesimal 16th Hilbert problem in the quadratic case*, Invent. Math. **143** (2001), no. 3, 449–497.

[GJ] F Girard and M.-A. Jebrane, *Majorations affines due nombre de zéros d'intégrales abéliennes pour les hamiltoniens quartiques elliptiques*, Ann. Fac. Sci. Toulouse Math. (6) **7** (1998), no. 4, 671–685.

[G] A. Glutsuk, *An explicit formula for the determinant of the Abelian integral matrix* (April 2000), available at `arXiv:math.DS/0004040`.

[GK] I. C. Gohberg and M. G. Kreĭn, *Systems of integral equations on the half-line with kernels depending on the difference of the arguments*, Uspehi Mat. Nauk (N.S.) **13** (1958), no. 2(80), 3–72 (Russian); English transl., Amer. Math. Soc. Transl. (2) **14** (1960), 217–287.

[H1] J. Harris, *Algebraic geometry*, Springer, New York, 1995. corrected reprint of the 1992 original.

[H2] P. Hartman, *Ordinary differential equations*, Birkhäuser, Boston, MA, 1982. reprint of the 2nd edition.

[H3] J. Heintz, *Definability and fast quantifier elimination in algebraically closed fields.*, Theoret. Comput. Sci. **24** (1983), no. 3, 239–277; corrigendum **39** (1985), no. 2-3, 343.

[H4] D. Hilbert, *Mathematical problems.*, Bull. Amer. Math. Soc. **8** (1902), 437–479; reprinted, Bull. Amer. Math. Soc. (N.S.) **37** (2000), no. 4, 407–436.

[HI] E. Horozov and I. D. Iliev, *Linear estimate for the number of zeros of Abelian integrals with cubic Hamiltonians*, Nonlinearity **11** (1998), no. 6, 1521–1537.

[I1] I. D. Iliev, *Higher-order Melnikov functions for degenerate cubic Hamiltonians*, Adv. Differential Equations **1** (1996), no. 4, 689–708.

[I2] Yu. S. Ilyashenko, *Appearance of limit cycles by perturbation of the equation $dw/dz = -R_z/R_w$, where $R(z,w)$ is a polynomial*, Mat. Sb. (N.S.) **78(120)** (1969), no. 3, 360–373 (Russian).

[I3] ———, *The multiplicity of limit cycles arising by a perturbation of a Hamilton equation of the class $dw/dz = P_2/Q_1$, in real and complex domains*, Tr. Semin. im. I. G. Petrovskogo **3** (1978), 49–60 (Russian).

[I4] ———, *Finiteness theorems for limit cycles*, Transl. Math. Monogr., vol. 94, Amer. Math. Soc., Providence, RI, 1991.

[I5] ———, *Bifurcation of planar and spatial polycycles: Arnold's program and its development*, The Arnoldfest (Toronto, 1997), Fields Inst. Commun., vol. 24, Amer. Math. Soc., Providence, RI, 1999, pp. 241–271.

[IY1] Yu. Ilyashenko and S. Yakovenko, *Double exponential estimate for the number of zeros of complete abelian integrals and rational envelopes of linear ordinary differential equations with an irreducible monodromy group*, Invent. Math. **121** (1995), no. 3, 613–650.

[IY2] ———, *Finite cyclicity of elementary polycycles in generic families*, Concerning the Hilbert 16th Problem, Amer. Math. Soc. Transl. Ser. 2, vol. 165, Amer. Math. Soc., Providence, RI, 1995, pp. 21–95.

[IY3] ———, *Counting real zeros of analytic functions satisfying linear ordinary differential equations*, J. Differential Equations **126** (1996), no. 1, 87–105.

[K1] A. G. Khovanskiĭ, *Real analytic manifolds with the property of finiteness, and complex abelian integrals*, Funktsional. Anal. i Prilozhen. **18** (1984), no. 2, 40–50 (Russian); English transl., Functional Anal. Appl. **18** (1984), no. 2, 119–127.

[K2] ———, *Fewnomials*, Transl. Math. Monogr., vol. 88, Amer. Math. Soc., Providence, RI, 1991.

[KY] A. G. Khovanskiĭ and S. Yakovenko, *Generalized Rolle theorem in $\mathbb{R}^n$ and $\mathbb{C}$*, J. Dynam. Control Systems **2** (1996), no. 1, 103–123.

[K] V. P. Kostov, *Fuchsian linear systems on $\mathbb{C}P^1$ and the Riemann–Hilbert problem*, C. R. Acad. Sci. Paris Sér. I Math. **315** (1992), no. 2, 143–148.

[KS] A. Kotova and V. Stanzo, *On few-parameter generic families of vector fields on the two-dimensional sphere*, Concerning the Hilbert 16th Problem, Amer. Math. Soc. Transl. Ser. 2, vol. 165, Amer. Math. Soc., Providence, RI, 1995, pp. 155–201.

[L] A. Ju. Levin, *The non-oscillation of solutions of the equation* $x^{(n)} + p_1(t)x^{(n-1)} + \cdots + p_n(t)x = 0$, Uspehi Mat. Nauk **24** (1969), no. 2(146), 43–96 (Russian); English transl., Russian Math. Surveys **24** (1969), no. 2, 43–99.

[Ł] S. Łojasiewicz, *Introduction to complex analytic geometry*, Birkhäuser Verlag, Basel, 1991.

[M] P. Mardešić, *An explicit bound for the multiplicity of zeros of generic Abelian integrals*, Nonlinearity **4** (1991), no. 3, 845–852.

[MS1] G. Moreno Socías, *An Ackermannian polynomial ideal*, Applied Algebra, Algebraic Algorithms and Error-Correcting Codes (New Orleans, LA, 1991), Lecture Notes in Comput. Sci., vol. 539, Springer, Berlin, 1991, pp. 269–280.

[MS2] ______, *Length of polynomial ascending chains and primitive recursiveness*, Math. Scand. **71** (1992), no. 2, 181–205.

[N1] D. Novikov, *Systems of linear ordinary differential equations with bounded coefficients may have very oscillating solutions*, Proc. Amer. Math. Soc. **129** (2001), no. 12, 3753–3755, electronic.

[N2] ______, *Modules of the Abelian integrals and the Picard–Fuchs systems*, Nonlinearity **15** (2002), no. 5, 1435–1444.

[NY1] D. Novikov and S. Yakovenko, *Simple exponential estimate for the number of real zeros of complete Abelian integrals*, Ann. Inst. Fourier (Grenoble) **45** (1995), no. 4, 897–927.

[NY2] ______, *Meandering of trajectories of polynomial vector fields in the affine n-space*, Publ. Mat. **41** (1997), no. 1, 223–242.

[NY3] ______, *Tangential Hilbert problem for perturbations of hyperelliptic Hamiltonian systems*, Electron. Res. Announc. Amer. Math. Soc. **5** (1999), 55–65, electronic.

[NY4] ______, *Trajectories of polynomial vector fields and ascending chains of polynomial ideals*, Ann. Inst. Fourier (Grenoble) **49** (1999), no. 2, 563–609.

[NY5] ______, *Redundant Picard–Fuchs system for Abelian integrals*, J. Differential Equations **177** (2001), no. 2, 267–306.

[NY6] ______, *Quasialgebraicity of Picard–Vessiot fields*, Moscow Math. J. **3**, no. 2, 551–591.

[P1] G. S. Petrov, *Elliptic integrals and their nonoscillation*, Funktsional. Anal. i Prilozhen. **20** (1986), no. 1, 46–49 (Russian).

[P2] ______, *The Chebyshev property of elliptic integrals*, Funktsional. Anal. i Prilozhen. **22** (1988), no. 1, 83–84 (Russian); English transl., Funct. Anal. Appl. **22** (1988), no. 1, 72–73.

[P3] ______, *Nonoscillation of elliptic integrals*, Funktsional. Anal. i Prilozhen. **24** (1990), no. 3, 45–50 (Russian); English transl., Funct. Anal. Appl. **24** (1990), no. 3, 205–210.

[P4] J. Plemelj, *Problems in the sense of Riemann and Klein*, Interscience Tracts Pure Appl. Math., vol. 16, Interscience Publishers John Wiley & Sons Inc., New York–London–Sydney, 1964.

[R1] R. Roussarie, *A note on finite cyclicity property and Hilbert's 16th problem*, Dynamical Systems (Valparaiso, 1986), Lecture Notes in Math., vol. 1331, Springer, Berlin, 1988, pp. 161–168.

[R2] ______, *Cyclicité finie des lacets et des points cuspidaux*, Nonlinearity **2** (1989), no. 1, 73–117.

[R3] ______, *Bifurcation of planar vector fields and Hilbert's sixteenth problem*, Progr. Math., vol. 164, Birkhäuser Verlag, Basel, 1998.

[RY] M. Roitmanm and S. Yakovenko, *On the number of zeros of analytic functions in a neighborhood of a Fuchsian singular point with real spectrum*, Math. Res. Lett. **3** (1996), no. 3, 359–371.

[dlVP] C. J. de la Vallée Poussin, *Sur l'équation différentielle linéaire du second ordre. Détermination d'une intégrale par deux valeurs assignées. extension aux équations d'ordre n*, J. Math. Pures Appl. **8** (1929), 125–144.

[V] A. N. Varchenko, *Estimation of the number of zeros of an abelian integral depending on a parameter, and limit cycles*, Funktsional. Anal. i Prilozhen. **18** (1984), no. 2, 14–25 (Russian); English transl., Functional Anal. Appl. **18** (1984), no. 2, 98–108.

[Y1] S. Yakovenko, *A geometric proof of the Bautin theorem*, Concerning the Hilbert 16th Problem, Amer. Math. Soc. Transl. Ser. 2, vol. 165, Amer. Math. Soc., Providence, RI, 1995, pp. 203–219.

[Y2] ———, *On functions and curves defined by ordinary differential equations*, The Arnoldfest (Toronto, 1997), Fields Inst. Commun., vol. 24, Amer. Math. Soc., Providence, RI, 1999, pp. 203–219.

[Y3] ———, *On zeros of functions from Bernstein classes*, Nonlinearity **13** (2000), no. 4, 1087–1094.

[Y4] Y. Yomdin, *Oscillation of analytic curves*, Proc. Amer. Math. Soc. **126** (1998), no. 2, 357–364.

CHAPTER 4

Around the Hilbert – Arnol′d Problem

V. KALOSHIN

1. Around the Hilbert's 16th problem and an estimate for cyclicity of elementary polycycles

1.1. The Hilbert 16th problem and its offsprings. Consider a polynomial vector field on the real (x, y)-plane

$$(1) \qquad \begin{cases} \dot{y} = P_n(x,y) \\ \dot{x} = Q_n(x,y) \end{cases} \qquad P_n, Q_n \text{ polynomials, } \deg P_n, \deg Q_n \leq n.$$

A limit cycle of a polynomial vector field (1) is an isolated periodic solution. Define

$$H(n) = \boxed{\text{uniform bound for the number of limit cycles of (1)}}.$$

One way to formulate the Hilbert 16th problem is the following:

HILBERT 16TH PROBLEM (HP). *Estimate $H(n)$ for any $n \in \mathbb{Z}_+$.*

To prove that $H(1) = 0$ is an exercise, but to find $H(2)$ is already a difficult unsolved problem (see [**DRR, DMR**] for work in this direction). Below we discuss two of the most significant branches of research HP has generated: existential and tangential Hilbert 16th problems.

1.1.1. *The tangential Hilbert 16th problem.* Consider a polynomial perturbation of a Hamiltonian polynomial line field

$$(2) \qquad \begin{cases} \dot{x} = -\dfrac{\partial H}{\partial x} - \varepsilon P(x,y) \\ \dot{y} = \dfrac{\partial H}{\partial y} - \varepsilon Q(x,y). \end{cases}$$

For $\varepsilon = 0$ the line field (2) does not have any limit cycles at all (all cycles are nonisolated). An oval (topological circle) γ of the level curve $H(x,y) = h$ generates a limit cycle for a small nonzero value of ε if the accumulated energy dissipation is zero in the first approximation, i.e., when

$$(3) \qquad \oint_\gamma P(x,y)\,dx + Q(x,y)\,dy = 0, \quad \gamma \subseteq \{H(x,y) = h\}.$$

The left-hand side expression is called a *complete Abelian integral.* If the polynomials H, P, and Q are fixed, the integral (3) defines a multivalued function $I(h)$. The multivalued possibility appears when the corresponding level curve $\{H(x,y) = h\}$ has several disjoint ovals.

TANGENTIAL HILBERT 16TH PROBLEM (THP) ([**AVI**$^+$]). *For any collection of polynomials P and $Q \in \mathbb{R}[x,y]$ of degree $\le n$ and H of degree $\le n+1$ give an upper bound* TH(n) *on the number of real ovals over which the integral* (3) *vanishes, but not identically.*

In the latter case the perturbation (2) is a Hamiltonian system for $\varepsilon \neq 0$ so it does not have any limit cycles at all. Even though the tangential Hilbert problem is not yet solved, in contrast to the Hilbert 16th problem, there are several quite general results related to it. Khovanskiĭ [**K1**] and Varchenko [**V**] proved

FINITENESS THEOREM. *For any $n \in \mathbb{Z}_+$ the number of isolated zeroes of Abelian integrals is uniformly bounded over all Hamiltonians and forms of degree $\le n$.*

For various other results estimating $H(n)$ in various particular cases see [**G, I1, M1, NDY, P2**], and the lecture course [**Y1**] in the present volume for more references.

If we consider (3) over the field of complex numbers we have that an Abelian integral satisfies a Fuchsian equation or Picard–Fuchs equation (see, e.g., [**AA**]), i.e., an equation of the form

$$\dot{z} = \sum_j \frac{A_j}{t-\alpha_j} z_j, \tag{4}$$

where the A_j are constant matrices and $z = (z_1, \dots, z_p) \in \mathbb{C}^p$ is a complex vector for some p. Investigation of various properties of fuchsian equations is the main topic of lectures of Bolibrukh [**B1**] in the present volume.

1.1.2. *From existential Hilbert* 16*th problem to Hilbert–Arnol'd problem.* A qualitative form of Hilbert 16th problem is the following:

EXISTENTIAL HILBERT 16TH PROBLEM (EHP). *Prove that $H(n) < \infty$ for any $n \in \mathbb{Z}_+$.*

The problem of the finiteness of the number of limit cycles for an individual polynomial line field (1) is called the *Dulac problem* after the pioneering work of Dulac [**D1**] who claimed in 1923 to have solved this problem, but an error was found by Ilyashenko [**I2**] 60 years later. The Dulac problem was solved by two independent, rather different and incredibly complicated proofs given almost simultaneously by Ilyashenko and Écalle.

INDIVIDUAL FINITENESS THEOREM (IFT) ([**I3, É**]). *Any polynomial line field* (1) *has only a finite number of limit cycles.*

However, neither proof allows any generalization to solve EHP. Consider the equation (1) for different polynomials $\big(P_n(x,y), Q_n(x,y)\big) \in (\mathbb{R}[x,y])^2$ as the family of vector fields on $\mathbb{R}^2$ depending on the parameters of the polynomials. Using a central projection $\pi\colon \mathbb{S}^2 \to \mathbb{R}^2$ and homogeneity with respect to parameters of the equation (1) (vector fields $\big(\lambda P_n(x,y), \lambda Q_n(x,y)\big)$ and $\big(P_n(x,y), Q_n(x,y)\big)$ for any $\lambda \neq 0$ have the same trajectories) one can construct a *finite parameter family of analytic line fields on the sphere $\mathbb{S}^2$ with a compact parameter base B* (see, e.g., [**IY2**] for details). After this reduction the existential Hilbert problem becomes a particular case of the following:

GLOBAL FINITENESS CONJECTURE (GFC) ([**R1**]). *For any analytic family of line fields on $\mathbb{S}^2$ with a compact parameter base B the number of limit cycles is uniformly bounded over all parameter values.*

We refer the reader to the volumes [**IY2, S1**], and a book [**R2**], where various developments of these and related problems are discussed. Families of analytic fields are difficult to analyze. In the middle of the 1980's Arnol′d [**AA**] proposed to consider generic families of smooth vector fields as the first step toward understanding families of analytic vector fields. A smooth analog of the global finiteness conjecture is the following:

HILBERT – ARNOL′D PROBLEM (HAP) ([**I4**]). *Prove that, in a generic finite parameter family of vector fields on the sphere S^2 with a compact base B, the number of limit cycles is uniformly bounded.*

Assume for a moment that an analytic (or a generic smooth) vector field on the sphere $\mathbb{S}^2$ has an infinite number of limit cycles. By the Poincaré – Bendixson theorem, any limit cycle should surround an equilibrium point and, since our vector field has at most finitely many equilibria, there should be an infinite "nested" sequence around one of the equilibria. Then this "nested" sequence of limit cycles has to accumulate (in the sense of the Hausdorff metric) to a certain contour (polygon) consisting of equilibria (as vertices) and separatrix curves (sides of that polygon) connecting them. Such objects are called *polycycles.* It turns out that a possible solution to the Hilbert – Arnol′d problem reduces to the investigation of bifurcation of polycycles. Let us give several definitions.

DEFINITION 1. A polycycle γ of a vector field on the sphere $\mathbb{S}^2$ is a cyclically ordered collection of equilibrium points $p_1, \dots, p_k$ (with possible repetitions) and arcs $\gamma_1, \dots, \gamma_k$ (distinct integral curves consisting possibly of equilibrium points) connecting them in a specific order: the jth arc γ_j connects p_j with p_{j+1} for $j = 1, \dots, k$.

A polycycle γ is called monodromic if one can choose a segment Σ transversal to γ such that on one side $U \subset \Sigma$ of $p = \Sigma \cap \gamma$ a Poincaré return map $\Delta_\gamma \colon U \to \Sigma$ is defined with $\Delta_\gamma(p) = p$.

NONACCUMULATION THEOREM ([**I3, É**]). *For any analytic monodromic polycycle γ there is a tube neighborhood free from limit cycles or a Poincaré return map $\Delta_\gamma \colon \Sigma \supset U \to \Sigma$ can't have infinitely many fixed points accumulating to $p = \Sigma \cap \gamma$.*

This theorem along with the above compactness arguments implies IFT. Both proofs of Ilyashenko and Ecallé deal with the analysis of type of germs of maps arising as Poincaré return maps of an analytic monodromic polycycle. The lectures by van den Dries discussed the theory of o-minimality.[1] This theory deals with classes of functions which satisfy certain axioms. A basic example of o-minimal class of functions is polynomials and analytic functions. In particular, if a map of a compact interval $\Delta \colon U \to \mathbb{R}$ belongs to an o-minimal class of functions, then the equation $\Delta(x) = x$ has finitely many solutions. One hopes that deeper understanding of o-minimal structures would allow one to include Poincaré return maps of monodromic polycycles into an o-minimal class and give an independent proof of

[1]For information on o-minimality and on model theory and its applications cf. [**vdD, vdDS**] and Lou van Dries webpage `http://www.math.uiuc.edu/People/vddnest.html`.

the nonaccumulation theorem. Finiteness theorems for differentiable function fields are discussed in the lectures by Buium [**B2**] in the present volume.

DEFINITION 2. Let $\{\dot{x} = v(x,\varepsilon)\}_{\varepsilon\in B^k}$, $x \in \mathbb{S}^2$, be a k-parameter family of vector fields on $\mathbb{S}^2$ having a polycycle γ for some parameter value $\varepsilon_* \in B^k$, where B^k is the unit ball in $\mathbb{R}^n$. The polycycle γ has cyclicity μ in the family $\{v(x,\varepsilon)\}_{\varepsilon\in B^k}$ if there exist neighborhoods U and V such that $\mathbb{S}^2 \supseteq U \supset \gamma$, $B^k \supseteq V \ni \varepsilon_*$ and for any $\varepsilon \in V$ the field $v(\cdot,\varepsilon)$ has no more than μ limit cycles inside U and μ is the minimal number with this property.

EXAMPLES. (1) In a generic n-parameter family, the maximal multiplicity of a degenerate limit cycle does not exceed $n+1$, e.g., in codimension 1 a semistable limit cycle has cyclicity 2. Thus, the cyclicity of a trivial polycycle (a polycycle without singular points) in a generic n-parameter family does not exceed $n+1$.

(2) (Andronov–Leontovich, 1930s; Hopf, 1940s). A nontrivial polycycle of codimension 1 has cyclicity at most 1.

(3) (Takens, Bogdanov, Leontovich, Mourtada, Grozovskii, early 1970s–1993 (see [**G2, KS**], and references there)). A nontrivial polycycle of codimension 2 has cyclicity at most 2.

There is no standard definition of a polycycle. For example in [**DRR**] the authors use the following definition: A *graphic* is a collection of singularities connected by phases curves. A polycycle is a graphic for which the first Poincaré return map is defined.

DEFINITION 3. The bifurcation number $B(k)$ is the maximal cyclicity of a nontrivial polycycle occurring in a generic k-parameter family of C^∞ vector fields.

The definition of $B(k)$ does not depend on a choice of the base of the family, it depends only on the number k of parameters.

LOCAL HILBERT–ARNOL'D PROBLEM (LHAP) ([**I4**]). *Prove that for any finite k, the bifurcation number $B(k)$ is finite and find an upper estimate for $B(k)$.*

It turns out that a solution to the local Hilbert–Arnol'd problem implies a solution to the Hilbert–Arnol'd problem.

Similarly to the generic smooth vector fields, in the case of analytic vector fields one can define a so-called *limit periodic set* [**FP, R1**], which is either a polycycle or has an arc of equilibrium points,[2] and formulate the

LOCAL FINITENESS CONJECTURE (LFC) ([**R1**]). *Prove that any limit periodic set occurring in an analytic family of vector fields on $\mathbb{S}^2$ has finite cyclicity in this family.*

Smooth vector fields are more flexible then analytic vector fields and easier to analyze. A strategy to attack the existential Hilbert problem, proposed by Arnol'd [**AA**] (see also [**IK**]), is first to understand generic smooth vector fields and then try to apply the developed methods to analytic vector fields. We stress out that a solution to the (local) Hilbert–Arnol'd problem would not necessarily solve the (finite cyclicity) global finiteness conjecture, because the study of generic vector fields does not include the study of nongeneric vector fields and nongeneric limit sets, which are unavoidable in the analytic setting. However, it might give some insight. Let us summarize the discussion in the form of a part of Figure 1.

[2] Generic vector fields cannot have an arc of equilibrium points.

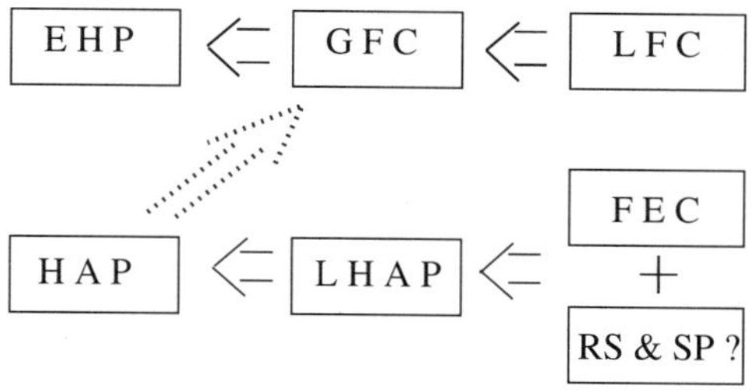

FIGURE 1. Existential Hilbert problem and its offsprings

1.1.3. *Cyclicity of elementary polycycles.* Now we shall formulate the main result of this course of lectures.

DEFINITION 4. An equilibrium point of a vector field on the two-sphere is called *elementary* if at least one eigenvalue of its linear part is nonzero. A polycycle is called an *elementary* polycycle if all its singularities are elementary.

The local Hilbert – Arnol′d problem was solved under the additional assumption that a polycycle have only elementary singularities.

DEFINITION 5. The elementary bifurcation number $E(k)$ is the maximal cyclicity of a nontrivial elementary polycycle occurring in a generic k-parameter family of C^∞ vector fields.

From Examples (2) and (3) above it follows that

$$E(1) = 1, \quad E(2) = 2.$$

Information about the behavior of the function $k \mapsto E(k)$ has been obtained recently. The first crucial step was done by Ilyashenko and Yakovenko:

FINITENESS OF ELEMENTARY CYCLICITY (FEC) ([**IY2**]). *For any n the elementary bifurcation number $E(n)$ is finite.*

COROLLARY 1. *Under the assumption that generic C^∞ families of vector fields have only elementary singularities, the global Hilbert – Arnol′d problem is solved, i.e., any generic finite-parameter C^∞ family of vector fields on the sphere $\mathbb{S}^2$ with a compact base and only elementary equilibria has a uniform upper bound for the number of limit cycles.*

MAIN THEOREM ([**K6**]). *For any $k \in \mathbb{Z}_+$*

$$E(k) \leq 2^{25k^2}. \tag{5}$$

This is the first explicit general estimate for cyclicity of polycycles. The case of a polycycle consisting of only one singular point with no arcs at all is well known. An elementary equilibrium point can generate limit cycles in its small neighborhood if it is a slow focus, that is, the linearization matrix has a pair of two imaginary eigenvalues. This bifurcation was investigated by Takens [**T1**].

COROLLARY 2. *As in Corollary 1, under the assumption that all the polycycles are elementary the Main Theorem gives a solution to the local Hilbert – Arnol′d problem.*

1.1.4. *Resolution of singularities* (RS) *or blow-up of singularities of vector fields and singular perturbations* (SP). In this section we discuss resolution of singularities and singular perturbations which might lead to a generalization of the main result and to a solution to the local Hilbert–Arnol'd problem (see the box with **RS & SP?** in Figure 1).

Let $\dot{x} = v(x)$ be a C^∞ vector field on $\mathbb{R}^2$ such that $v(0) = 0$. A vector field satisfies a *Lojasiewicz condition* if there exists $k \in \mathbb{Z}_+$ and $c > 0$ such that

$$\|v(x)\| \geq c\|x\|^k \tag{6}$$

for all x in some neighborhood of 0. It can be shown [**D2**] that any generic C^∞ finite-parameter family of vector fields on the sphere $\mathbb{S}^2$ has only vector fields with equilibrium points satisfying a Lojasiewicz condition for some $k \in \mathbb{Z}_+$ and $c > 0$.

To define a *blow-up* for a C^∞ vector field $\dot{x} = v(x)$ on $\mathbb{R}^2$ with an equilibrium at 0, i.e., $v(0) = 0$ consider the map

$$\phi\colon \mathbb{S}^1 \times \mathbb{R} \to \mathbb{R}^2; \quad \phi(\theta, r) \mapsto (r\cos\theta, r\sin\theta). \tag{7}$$

Then the pull-back $\hat{v}$, with $\phi(\hat{v}) = v$, is a C^∞ vector field on $\mathbb{S}^1 \times \mathbb{R}$, i.e., $d\phi_0(\hat{v}(0)) = X \circ \phi(0)$, where $\hat{v}$ is the blown-up vector field.

DESINGULARIZATION THEOREM ([**D2**]). *If a C^∞ vector field $\dot{x} = v(x)$ on $\mathbb{R}^2$ with $v(0) = 0$ satisfies a Lojasiewicz condition, then there is a finite sequence of blow-ups leading to a vector field with only elementary equilibria.*

Sometimes this theorem is called the Bendixson–Seidenberg–Dumortier theorem [**B, S2, D2**]. Bendixson stated it without a proof it on the brink of the twentieth century. Seidenberg proved it in the analytic case and Dumortier did it for C^∞ vector fields with a Łojasiewicz condition. A quantitative version of the desingularization theorem, which estimates the number of necessary blow-ups, was obtained by Kleban [**K3**].

This theorem reduces consideration of an *individual* vector field, occurring in a generic finite-parameter family, with equilibria without restriction to an individual vector field with only *elementary* equilibria.

However, in order to extend an estimate on cyclicity of elementary polycycles (5) to an estimate on cyclicity of a generic nonelementary polycycle (LHAP) one needs the *desingularization theorem for families* of generic C^∞ vector fields. Different approaches to attack this problem were proposed by Denkowska and Roussarie [**DR**] and by Trifonov [**T4**].

An approach proposed by Trifonov leads to the dynamical phenomenon called *Singular Perturbation* (SP): in the simplest case one needs to analyze families of vector fields on the plane which for some values of the parameters have a curve of equilibria. Certainly a generic finite-parameter family of vector fields has no curve of equilibria; however, after even one step of blow-up such a curve can occur [**T4**]. Appearance of curves of equilibria after a desingularization in a family now seems to be the main obstacle standing between an estimate on cyclicity of elementary polycycles (5) and the local Hilbert–Arnol'd problem (see [**T4, IY2, R2**] for more).

1.2. Bifurcation of spatial polycycles and multiplicity of generic germs. In this part we present by-product results of the Main Theorem. The first result is an extension of the Main Theorem on the estimate of cyclicity of a planar elementary polycycle to an estimate on cyclicity of a spatial quasielementary polycycle (see Section 1.2.1). The second result gives an estimate on cyclicity of generic germs of smooth mappings which is a partial answer to Arnol′d's question [**A**] (see Section 1.2.4).

1.2.1. *Bifurcation of spatial polycycles.* The definition of a polycycle in the multidimensional case is the same as in the planar case (see Definition 1). When Ilyashenko and Yakovenko proved the Finiteness of Cyclicity for elementary polycycles, Arnol′d posed the question: *What can be said about bifurcations of spatial polycycles?*

Another sufficient reason to look at this problem is because the planar argument (the Poincaré – Bendixson theorem) implies that a collection of an infinite number of limit cycles of uniformly bounded length, located in a bounded domain, accumulates to a limit cycle. Indeed, consider a vector field $\dot{x} = v(x)$ of finite codimension in $\mathbb{R}^3$ (dimension 3 can be replaced by any $N > 2$ anywhere in this section), i.e., a vector field which occurs in a generic C^∞ finite-parameter family. Then $v(x)$ has only isolated singular points. Fix a positive number L and assume that in a compact region of the phase space there is an infinite number of phase curves of length less than L corresponding to limit cycles of $v(x)$. Then a subset of these limit cycles must accumulate to a separatrix polygon (polycycle).

Bifurcation properties of spatial polycycles are much richer then those of planar polycycles. The first important 3-dimensional feature is the existence of limit cycles that wind several times around a polycycle. This happens because a Poincaré return map is a 2-dimensional map and it might have not only fixed points, but also periodic points of higher periods as well. We call a periodic trajectory that "turns" around a whole polycycle exactly m times before closing up an *m-cycle.* Such a trajectory corresponds to a periodic point of a corresponding Poincaré return map of minimal period m. On the plane, for topological reasons, only 1-cycles exist. The definition of cyclicity requires some additional care.

Consider an n-parameter family of flows $\{\dot{x} = v(x, \varepsilon)\}_{\varepsilon \in B^n}$ in $\mathbb{R}^3$. Let $\gamma \subset \mathbb{R}^3$ be a polycycle of the field $\dot{x} = v(x, \varepsilon^*)$ for some $\varepsilon^* \in B^n$. Then γ can be represented as a union of a finite number of equilibrium points $\{p_j\}_{j \in J}$ and phase curves $\{\gamma_j\}_{j \in J}$ connecting them. A *tube neighborhood* T_γ of the polycycle γ is a union of neighborhoods of equilibria $\{p_j\}_{j \in J}$ and tube neighborhoods $\{T_j\}_{j \in J}$ of phase curves $\{\gamma_j\}_{j \in J}$.

Definition 6. Let $m \in \mathbb{Z}_+$. Then m-cyclicity of the polycycle γ in the family $\{\dot{x} = v(x, \varepsilon)\}_{\varepsilon \in B^n}$, denoted by $\mu(m, \gamma)$, is the minimal number $\mu(m, \gamma)$ for which there is a tube neighborhood T_γ of the polycycle $\gamma \subset T_\gamma \subset \mathbb{R}^3$, a neighborhood V of the parameter $\varepsilon^* \in V \subset B^n$ and for each $j \in J$ a Poincaré section $L_{\gamma,j}$ (a hyperplane) transversal to the corresponding γ_j such that the following condition holds:

(1) for any parameter $\varepsilon \in V$ the corresponding vector field $\dot{x} = v(x, \varepsilon)$ has at most $\mu(m, \gamma)$ limit cycles in T_γ;
(2) Each of these limit cycles $l_i(\varepsilon)$ intersects each Poincaré section $L_{\gamma,j}$ in exactly m different points;
(3) In the sense of the Hausdorff metric for each $j \in J$ the distance between each part of $l_j(\varepsilon)$ which lies between two consecutive intersections of $L_{\gamma,j}$ and the polycycle γ tends to 0 as ε tends to 0.

Now we discuss a classical example of a polycycle which has infinite m-cyclicity for any $m \geq 1$.

1.2.2. *The Shil'nikov polycycle.* Consider a flow ϕ_t in $\mathbb{R}^3$ with a hyperbolic equilibrium point O that has one positive eigenvalue λ and two complex conjugates $\mu \pm \omega i$ with negative real part. Suppose that the sum of $\lambda + \mu$ is positive, and the unstable one-dimensional manifold returns to the stable one, which is two-dimensional. Thus, the equilibrium O has a homoclinic orbit that tends back to O along the unstable manifold as $t \to -\infty$, and spirals around O on the stable manifold as $t \to +\infty$. In 1965 Shil'nikov [**S3**] discovered that the Poincaré map along this polycycle has a countable number of pairwise disjoint subdomains so that a restriction to any of them gives a Smale horseshoe. Any such horseshoe is structurally stable, therefore, the polycycle described above (the Shil'nikov polycycle) has infinite m-cyclicity for all $m \in \mathbb{Z}_+$ (see, e.g., [**GH, IL**]). The codimension of this polycycle is 1.

However, it seems reasonable to state the following

CONJECTURE 1 (Arnol'd–Ilyashenko–Yakovenko). *If a spatial polycycle $\gamma \in \mathbb{R}^3$ has finite codimension k and all its equilibrium points are saddles with real eigenvalues or saddlenodes with at most one zero eigenvalue and the other eigenvalues are real, then the m-cyclicity of γ, denoted $C(m,\gamma)$, is finite for each $m \in \mathbb{Z}_+$.*

Using the ideas and methods for the planar problem and a result of Grigoriev–Yakovenko [**GY**], the Arnol'd–Ilyashenko–Yakovenko's conjecture has been solved in arbitrary dimension $N > 2$ with additional nondegeneracy assumptions on the polycycle's equilibria.

1.2.3. *An estimate of the cyclicity of a quasielementary spatial polycycle.* In the planar case we considered polycycles with elementary equilibria only, now we define a class of points called *quasielementary equilibria.* The author has shown that polycycles with only quasielementary equilibria, have finite m-cyclicity for any $m \geq 1$. Moreover, there exists an explicit upper bound for m-cyclicity.

Recall some standard definitions from normal form theory.

DEFINITION 7. The set of complex numbers $\lambda_1, \ldots, \lambda_N \in \mathbb{C}$ is called:

(a) nonresonant if there is no integral relation among the numbers λ_j of the form $\lambda_j = \sum_{i=1}^N k_i\lambda_i$, where $k_i \in \mathbb{Z}_+$ for $i = 1, \ldots, n$ and $\sum_{i=1}^N k_i \geq 2$.

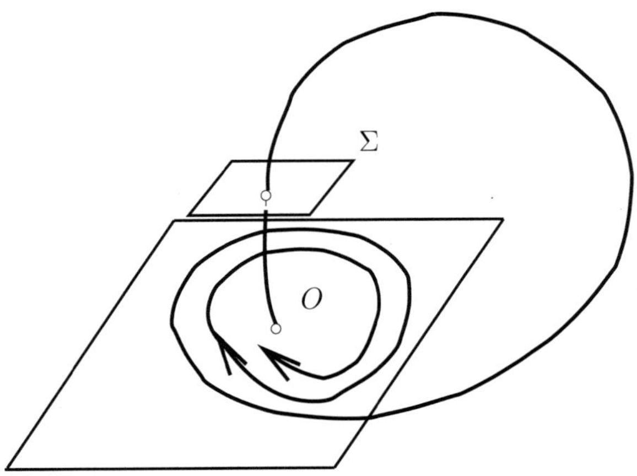

FIGURE 2. The Shil'nikov polycycle

(b) *strongly simply resonant* if all the nontrivial resonance relations $\lambda_j = \sum_{i=1}^{N} k_i \lambda_i$, where $k_i \in \mathbb{Z}_+$, $i = 1, \dots, n$ and $\sum_{i=1}^{N} k_i \geq 2$, follow from a single one $\sum_{i=1}^{N} k_i^* \lambda_i = 0$, where $k_i^* \in \mathbb{Z}_+$.

DEFINITION 8. We shall call an equilibrium point of a differential equation *quasielementary* if the linearization matrix of the equation at this point has only *real* eigenvalues, at most one of them is zero, and they satisfy one of the following conditions:

(1) they are nonresonant, and we call such an equilibrium a *nonresonant saddle*;
(2) they form a strongly simply resonant set of numbers — a *strongly simply resonant saddle*;
(3) one eigenvalue is zero with Łojasiewicz exponent 2 (see (6)) and the others form a nonresonant set.

A polycycle is called *quasielementary* if all its vertices are quasielementary.

Note that the class of quasielementary points in the case of the plane ($N = 2$) coincides with the class of elementary points, except for the multiplicity two condition for saddlenodes. In a sense, Theorem 1 below is a generalization of Theorem 5 of Section 2.2 below.

DEFINITION 9. The *quasielementary bifurcation number* $\mathrm{QE}(N, n, m)$ is the maximal m-cyclicity of a quasielementary polycycle occurring in a generic n-parameter family of vector fields in $\mathbb{R}^N$.

THEOREM 1 ([**K8**]). *For any positive integer N (dimension of the phase space), n (number of parameters), m (number of turns around a polycycle), and $T = 6Nnm$ we have*

$$\mathrm{QE}(N, n, m) \leq 2^{T^2}.$$

In the next section we describe another by-product of the Main Theorem.

1.2.4. *Geometric multiplicity of germs of generic maps.* Let $F \colon \mathbb{R}^n \to \mathbb{R}^n$ be a generic C^k smooth map, $k \geq n + 1$. Fix a point $a \in \mathbb{R}^n$ and denote $F(a)$ by b.

DEFINITION 10. The geometric multiplicity of a map germ $F \colon (\mathbb{R}^n, a) \to (\mathbb{R}^n, b)$ at a, denoted by $\mu_a^G = \mu_a^G(F)$, is the maximal number of isolated preimages $F^{-1}(\tilde{b})$ close to a:

$$\mu_a^G(F) = \limsup_{r \to 0} \sup_{\tilde{b} \in \mathbb{R}^n} \#\{x \in B_r(a) : F(x) = \tilde{b}\}. \tag{8}$$

For example, the geometric multiplicity of the function $f \colon x \to x^2$ at 0 is two, but the geometric multiplicity of $f \colon x \to x^3$ at 0 is one, even though 0 is a degenerate point of second order.

In the complex case the geometric multiplicity equals the usual multiplicity (see, e.g., [**AGV**]). In the real case the first is no greater than the second.

DEFINITION 11. Define the geometric multiplicity of n-dimensional germs, $\mu^G(n)$, as follows. Let $F \colon \mathbb{R}^n \to \mathbb{R}^n$ be a generic map. The geometric multiplicity of F equals the least upper bound of the geometric multiplicities of $\mu_a^G(F)$ taken over all points $a \in \mathbb{R}^n$. Then the geometric multiplicity of n-dimensional germs is the maximum of the geometric multiplicities of all generic maps F from $\mathbb{R}^n$ to $\mathbb{R}^n$

$$\mu^G(n) = \sup_{\substack{F\text{-generic} \\ a\in\mathbb{R}^n}} \mu_a^G(F). \tag{9}$$

It turns out that the geometric multiplicity of n-dimensional germs is finite for all positive integers n and depends only on the dimension n.

REMARK 1. For example, for $n = 2$ the Whitney theorem about maps of surfaces states that a generic C^2 map of two-dimensional manifolds $F\colon M^2 \to N^2$ can have only three different types of germs: one-to-one, a fold, and a pleat (see, e.g., [**AGV**]). This implies that $\mu^G(2) = 3$.

A natural problem posed by Arnol′d [**A**] is *to give estimates for the geometric multiplicity $\mu^G(n)$ of n-dimensional germs.*

In the case of complex analytic maps of $\mathbb{C}^n$ into $\mathbb{C}^n$ Gabrielov and Khovanskiĭ [**GAK**, Theorem 7] obtained an estimate on $\mu^G(n)$ of the type: $\mu^G(n) \leq n^n$. The upper bound for the geometric multiplicity for n-dimensional *smooth* germs of generic maps is given by

THEOREM 2 ([**K6**]). *The geometric multiplicity of germs of a generic C^k smooth map $F\colon \mathbb{R}^n \to \mathbb{R}^n$, $k > n$ admits the following upper estimate:*

$$\mu_a^G(F) \leq 2^{n(n-1)/2+1} n^n, \quad \forall a \in \mathbb{R}^n. \tag{10}$$

Using the same method one can prove

THEOREM 3 ([**K6**]). *Let $F\colon \mathbb{R}^n \to \mathbb{R}^N$ be a generic C^k smooth map with $k > n$, $N \geq n$ and $P\colon \mathbb{R}^N \to \mathbb{R}^n$ be a polynomial of degree d. Then the geometric multiplicity of germs of a chain map $P \circ F\colon \mathbb{R}^n \to \mathbb{R}^n$ admits the following upper estimate:*

$$\mu_a^G(P \circ F) \leq 2^{n(n-1)/2+1} (dn)^n, \quad \forall a \in \mathbb{R}^n. \tag{11}$$

An interesting feature of this theorem is that the geometric multiplicity does not depend on the dimension N of the intermediate space.

The problem of an upper estimate of geometric multiplicity of germs of generic smooth maps $F\colon (\mathbb{R}^n, 0) \to (\mathbb{R}^n, 0)$ or chain maps $P\circ F\colon (\mathbb{R}^n, 0) \to (\mathbb{R}^n, 0)$ is closely related to the problem of estimating the cyclicity of elementary polycycles as the reader will see below.

All the results (the Main Theorem, Theorem 1, Theorem 2, and Theorem 3) were announced in [**K6**].

1.3. Three stages of the proof of the Main Theorem and outline of the content of the lectures. The Main Theorem is a quantitative extension of the Ilyashenko–Yakovenko finiteness theorem. The paper of Ilyashenko–Yakovenko [**IY2**] was a cornerstone for the present paper. In [**IY2**] the authors made an important step: they found a route from *bifurcation theory* to *singularity theory* using the Khovanskiĭ reduction method [**K1**]. In [**K6**] we follow this route at the beginning and using some new ideas get an estimate for the cyclicity of elementary polycycles. Below we outline the main steps of the proof of the Main Theorem and describe the contents of the coming sections.

The proof of the Main Theorem consists of three steps. The relation to the proof of the finiteness theorem [**IY2**] is discussed in Section 1.3.2.

Step 1. *Normal forms for local families of vector fields and their integration.* We use normal forms to establish an explicit form for the Poincaré correspondence map near equilibrium points on the polycycle under consideration. In [**MR**] and later in [**IY2**] it is noticed that these maps satisfy Pfaffian (polynomial differential) equations with the coefficients of polynomials depending smoothly on the parameters of the family. As a result a *basic system* of equations, determining the number of limit cycles, is obtained.

Step 2. *The Khovanskiĭ reduction method.* We discuss a variation of the Khovanskiĭ method [**K2**]. This method allows us to investigate systems of equations that involve functions satisfying Pfaffian equations. It turns out that the number of solutions of the basic system can be estimated by the number of solutions of a *mixed functional-Pfaffian* system. After an application of the Khovanskiĭ method to the mixed functional-Pfaffian system we obtain several *chain maps*: the maps of the form

$$x \mapsto (P_1, \dots, P_n) \circ \big(x, f(x), f'(x), \dots, f^{(n)}(x)\big), \tag{12}$$

where x is a point near $0 \in \mathbb{R}^n$, f is a generic function, $f^{(k)}(x)$ is the collection of all derivatives of f of order k, and $P = (P_1, \dots, P_n)$ is a vector-polynomial given by its coordinate functions of known degree.

It turns out that the problem of estimating the number of limit cycles reduces to estimating the number of small regular preimages of some *special* points of the chain map. Special points form an open cone-like semialgebraic set K approaching 0 in the image, e.g., if $K \subset \mathbb{R}^2$, then $K = \{(x_1, x_2) : 0 < x_2 < x_1^m\}$ for some $m \in \mathbb{Z}_+$.

Denote by F the map $F\colon x \mapsto \big(x, f(x), f'(x), \dots, f^{(n)}(x)\big)$ which is called the nth jet of f. Denote by L_F the linearization of F at the point $x = 0$.

Lecture 2 highlights Step 1 and in a simple nontrivial case.

Step 3. *Bézout's theorem for the Chain maps.* We shall construct an algebraic set Σ in the image of F (in the space of n-jets) so that if F is transversal to Σ, then the number of preimages of any point a from a set of special points K is *the same* for F and its linearization L_F at zero:

$$\#\{x : P \circ F(x) = a\} = \#\{x : P \circ L_F(x) = a\} \le \prod_{j=1}^{k} \deg P_j. \tag{13}$$

Since L_F is a linear map, one can apply Bézout's theorem to estimate the right-hand part of the equality. This observation completes the proof of the Main Theorem.

In order to prove existence of such a set Σ we need to apply stratification theory originating in works of Whitney [**W2**], Thom [**T2**], and Mather [**M2**]. More exactly, we need to prove the existence of the so-called $\mathfrak{a}_P$-stratification introduced by Thom in some special case [**K6**]. Lecture 3 presents necessary notions from stratification theory and states the required result on existence of an $\mathfrak{a}_P$-stratification.

1.3.1. *Multichain maps and bifurcation of spatial polycycles.* In order to get an estimate on cyclicity of spatial polycycles in Theorem 1 we face the problem of estimating the geometric multiplicity of *multichain maps* of the form

$$P \circ (F, F)\colon B^n \times B^n \to \mathbb{R}^{2n}, \tag{14}$$

where $B^n \subset \mathbb{R}^n$ is a unit ball, $F\colon B^n \to \mathbb{R}^N$ is a generic map and $P\colon \mathbb{R}^{2N} \to \mathbb{R}^{2n}$ is a vector-polynomial of known degree. The appearance of this problem is described

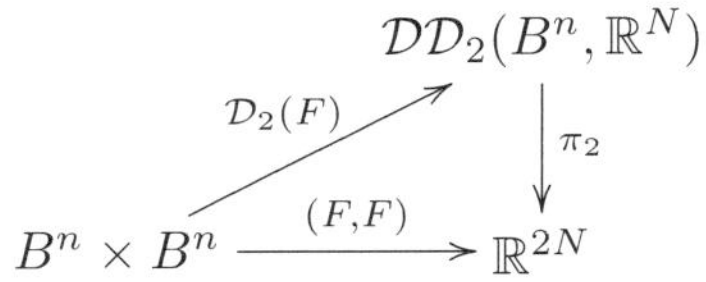

FIGURE 3. Polynomial blow-up of the multijet space

with many more details in Lecture 4. It is no longer possible to treat the 2-tuple map

$$(F, F)\colon B^n \times B^n \to \mathbb{R}^N \times \mathbb{R}^N \tag{15}$$

as a generic map.

Step 4. *Blow-up along the diagonal in the multijet space.* Grigoriev and Yakovenko [**GY**] constructed a so-called *space of divided differences or $\mathcal{DD}_2$-space* and the commutative diagram of Figure 3 where $\mathcal{D}_2(F)\colon B^n \times B^n \to \mathcal{DD}_2(B^n, \mathbb{R}^N)$ is a smooth map, $\pi_2\colon \mathcal{DD}_2(B^n, \mathbb{R}^N) \to \mathbb{R}^{2N}$ is an explicitly computable polynomial and

$$\pi_2 \circ \mathcal{D}_2(F) = (F, F) : B^n \times B^n \to \mathbb{R}^{2N}. \tag{16}$$

It turns out that one can treat $\mathcal{D}_2(F)$ as a generic map for a generic F and impose various transversality conditions. Therefore, we can represent the multichain map $P \circ (F, F)$, given by (14), in the form

$$P \circ (F, F) = (P \circ \pi_2) \circ \mathcal{D}_2(F), \tag{17}$$

where $P \circ \pi_2$ is a polynomial since π_2 is a polynomial, and $\mathcal{D}_2(F)$ is a smooth map. Moreover, it turns out that $\mathcal{D}_2(F)$ is generic for a generic F. Now we can apply Bézout's theorem to the chain map $(P \circ \pi_2) \circ \mathcal{D}_2(F)$. In Lecture 4 we describe this construction with details and in greater generality and exhibit application of this construction to an old problem on the rate of growth of the number of periodic points for generic diffeomorphisms in smooth dynamic systems (see, e.g., [**AM, S4**]).

1.3.2. *Relation of the proof of the Main Theorem and the Ilyashenko–Yakovenko finiteness theorem* [**IY2**]. Steps 1 of both proofs [**IY2**] and [**K6**] are the same. We shall just present the table of required normal forms from [**IY2**], which were obtained in [**IY1**]. Step 2 in this proof is slightly different from the one in [**IY2**] and this is the first novel point. After application of the Khovanskiĭ method we obtain the same collection of chain maps of the form (12) as in [**IY2**]. However, in [**IY2**] the authors investigate the number of regular preimages of points in the image by the chain maps *without any restriction* on those points. In the present proof, using new additional arguments in the Khovanskiĭ method, we reduce consideration to only preimages of *special* points, i.e., points from a tiny cone-like set in the image. At this point our proof continues on independently. The proof from [**K6**] can be considered as an independent simplified proof of the Ilyashenko–Yakovenko's finiteness theorem modulo the derivation of a mixed functional-Pfaffian system.

Acknowledgments. I would like to thank Dana Schlomiuk for giving me a great opportunity to give a course of lectures in a workshop held in Montréal during June 2000 and for careful reading of the present manuscript. Many helpful remarks improved its overall quality. Special thanks go to my teacher Yulij Ilyashenko who patiently taught me bifurcation theory and whose long lasting support and encouragement have been a crucial source of energy. Discussions with William

Cowieson, Andrei Gabrielov, Askold Khovanskiĭ, Pavao Mardešić, John Mather, Robert Moussu, Oleg Shelkovnikov, and Sergei Yakovenko have been very helpful for me. I would like also to thank Christiane Rousseau and Robert Roussarie for inviting me to give series of lectures on the bifurcation of limit cycles in Montréal and Dijon and for useful discussions. These series of lectures have been extremely helpful in improving the presentation of the present lectures. I would like to acknowledge financial support and the very productive atmosphere of IPST, University of Maryland and Courant Institute of Mathematical Science, NYU, where various parts of the work have been done. While in the Institute for Physical Sciences and Technology I enjoyed fruitful discussions with James Yorke and Brian Hunt.

2. Normal forms and the Khovanskiĭ method

We explain the proof of the Main Theorem in the simplest nontrivial case $n = 2$. Consider a generic 2-parameter C^∞ family of vector fields $\{\dot{x} = v(x, \varepsilon)\}_{\varepsilon \in B^2}$ and suppose that for $\varepsilon = 0$ the vector field $\dot{x} = v(x, 0)$ has a polycycle γ which consists of two saddles p_1, p_2 and two separatrices γ_1, γ_2 connecting them. Consider a segment Σ transversal to, say, γ_1 and denote by $\Delta\colon \Sigma \supset U \to \Sigma$ the Poincaré return map, which is defined on some open set U in Σ. In order to estimate the number of limit cycles bifurcating from the polycycle γ we need to *estimate the number of isolated fixed points* $\#\{x \in U : \Delta(x) = x\}$.

Using the standard approach we decompose the Poincaré map Δ into a composition of four maps: two local Δ_1 and Δ_2 in neighborhoods of equilibria p_1 and p_2 respectively and two semilocal f_1 and f_2 along connecting separatrices γ_1 and γ_2 respectively to be defined precisely below (see Figure 4). After that we replace the equation $\Delta(x) = x$ by the system of equations corresponding to $\Delta \equiv \Delta_2 \circ f_2 \circ \Delta_1 \circ f_1$. To understand properties of local maps Δ_i, $i = 1, 2$ we use normal form theory.

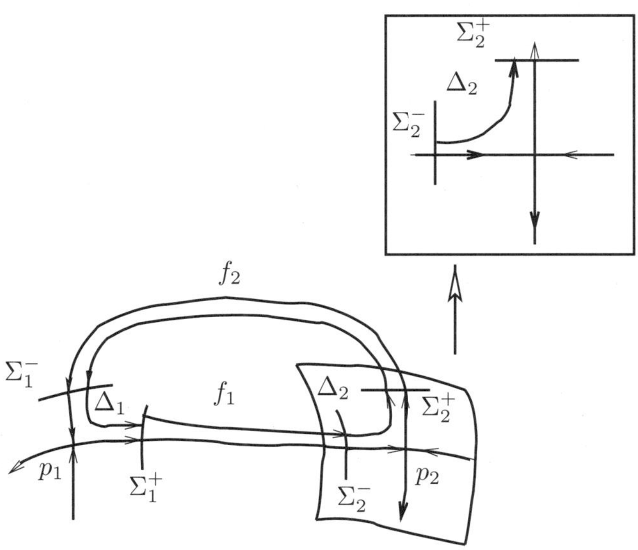

FIGURE 4. Construction of "entrance" and "exit" transversals

We say that a finite parameter family of vector fields is generic if generic finite parameter families form a dense set and pass through every point. It is essential that generic families pass through every point. This is somewhat similar to the notion of prevalence [**K1**].

2.1. Normal forms and a basic system determining the number of limit cycles.

2.1.1. *Polynomial normal forms of local families and Pfaffian Poincaré return maps.* It turns out that in a small neighborhood of an elementary equilibrium point there exist finitely differentiable normal coordinates (in the Cartesian product of the phase space and the parameter space), the so-called normal form of an equilibrium point. The list of finitely differentiable normal forms was obtained in [**IY1**]. The main feature of the list: all normal forms are *polynomial and integrable.* The smaller the neighborhood of a normal form, the higher is its smoothness. So smoothness can be chosen arbitrary large. All normal forms are summarized in Table 1 below.

In a small neighborhood of an elementary equilibrium point one can choose two small segments, say Σ^- and Σ^+, transversal to the vector field for the critical value of the parameter and explicitly calculate the Poincaré (correspondence) map which maps a point from one segment, say Σ^-, along the corresponding phase curve to a point from the other segment Σ^+ (see Figure 4). For an appropriate choice of segments Σ^-, Σ^+ and coordinate functions x, y in Σ^-, Σ^+ respectively, and a smooth function $\lambda(\varepsilon)$ in the original parameter ε of the family, the Poincaré return map $\Delta_\varepsilon\colon x \to y$, $\varepsilon \in \mathbb{R}^k$, can be explicitly computed. Moreover, there is a Pfaffian equation with 1-form ω of the form

$$P\big(x, y, \lambda(\varepsilon)\big)\, dx + Q\big(x, y, \lambda(\varepsilon)\big)\, dy = 0 \tag{18}$$

vanishing on the graph of $y = \Delta_\varepsilon(x)$, where $P\big(x, y, \lambda(\varepsilon)\big)$ and $Q\big(x, y, \lambda(\varepsilon)\big)$ are polynomials in x and y. This was first noticed by Moussu–Roche [**MR**].

EXAMPLE 1. Consider a nonresonant saddle on the plane in a family of C^∞ vector fields. There is a normal form which after an appropriate rescaling is given by the equation

$$\begin{cases} \dot{x} = \lambda_1(\varepsilon)x \\ \dot{y} = -\lambda_2(\varepsilon)y, \end{cases} \tag{19}$$

where $\lambda_1(\varepsilon)$ and $\lambda_2(\varepsilon)$ are smooth functions and two transversal "exit"-"entrance" sections are $\Sigma^- = \{y = 1\}$, $\Sigma^+ = \{x = 1\}$.

Then for $\lambda(\varepsilon) = \lambda_1(\varepsilon)/\lambda_2(\varepsilon)$ the function $u(t) = x(t)y^{\lambda(\varepsilon)}(t)$ is a first integral. Therefore, if a trajectory starts form $\big(x(0), y(0)\big) \in \Sigma^-$ and ends at $\big(x(t^*), y(t^*)\big) \in \Sigma^+$ (see Figure 5, S_μ-case), then $y^{\lambda(\varepsilon)}(t^*) = x(0)$ or in the induced on Σ^- and Σ^+ coordinates from $\mathbb{R}^2$ we have $y^{\lambda(\varepsilon)} = x$. It is easy to see that the 1-form $\omega = x\,dy + \lambda(\varepsilon)y\,dx$ vanishes on the graph $y^{\lambda(\varepsilon)} = x$. All necessary information about normal forms, Poincaré correspondence maps, and corresponding Pfaffian forms is in the Table 1 below. For completeness let's give the necessary definitions from the theory of normal forms.

2.1.2. *Definitions and a collection of normal forms.* A local family of planar vector fields is the germ of a map,

$$v\colon (\mathbb{R}^2, 0) \times (\mathbb{R}^k, 0) \to (\mathbb{R}^2, 0), \quad (x, y, \varepsilon) \mapsto v(x, y, \varepsilon).$$

TABLE 1. Unfolding of elementary equilibrium points on the plane

Type	Normal forms	Poincaré correspondence maps	Pfaffian equations
S_0	$\dot{x} = x,\ \dot{y} = -\lambda y,$ $\lambda = \lambda_0 \in \mathbb{R}^1$	$y = x^\lambda,$ $x > 0,\ y > 0$	$x\,dy - \lambda y\,dx = 0$
S_μ	$\dot{x} = x\big(n/m + P_\mu(u,\lambda)\big),$ $\dot{y} = -y,$ $u = u(x,y) = x^m y^n,$ $P_\mu(u,\lambda) = \pm u^\mu(1+\lambda_\mu u^\mu) + W_{\mu-1}(u,\lambda),$	$0 = m \log y + \int_{x^m}^{y^n} du/uP_\mu(u,\lambda),$ $x > 0,\ y > 0,$ $\lambda = (\lambda_1, \dots, \lambda_\mu)$	$yP_\mu(y^n,\lambda)\,dx - (n/m + P_\mu(y^n,\lambda)) \times xP_\mu(x^m,\lambda)\,dy = 0$
D^c_μ	$\dot{x} = Q_\mu(x,\lambda),$ $\dot{y} = -y,$ $Q_\mu(x,\lambda) = \pm x^{\mu+1}(1+\lambda_\mu x^\mu) +$	$y = C(\lambda)x,$ $C = \int_{-1}^{1} du/Q_\mu(u,\lambda),$ $x, y \in \mathbb{R}^1$	$x\,dy - y\,dx = 0$
D^h_μ	$W_{\mu-1}(x,\lambda),$ $\lambda = (\lambda_1, \dots, \lambda_\mu)$	$0 = \log y + \int_x^1 du/Q_\mu(u,\lambda),$ $y > 0,\ x \in \mathbb{R}^1$	$Q_\mu(x,\lambda)\,dy - y\,dx = 0$

In the first column we use the following notation for elementary equilibria (the subscript indicates the degree of degeneracy):

S_0 — Nonresonant saddle;
S_μ — Resonant saddle whose quotient equation (the differential equation for $u = x^m y^n$ below) has the singular point of multiplicity $\mu + 1$ at the origin, $\mu \geq 1$; if we want to specify explicitly the resonance between the eigenvalues, we use the extended notation $S_\mu^{(n:m)}$ assuming that the natural numbers m, n are mutually prime;
D_μ — Degenerate saddlenode of multiplicity μ;
$W_{\mu-1}(z,\lambda) = \lambda_0 + \lambda_1 z + \dots + \lambda_{\mu-1} z^{\mu-1}$ is a Weierstrass polynomial of degree $\mu - 1$.

For different technical remarks concerning this table see [**IY2**, §1.1]. We just briefly describe each column.
The second column has the corresponding normal forms. In the third column of the table the Poincaré correspondence maps $y = \Delta(x,\lambda)$ for the polynomial normal forms are given. They are implicitly defined by the equations relating x to y, these equations depending explicitly on the parameters λ and thus implicitly on the original parameters ε. The choice of segments transversal to the phase curves of the family is described in Figure 4. The last column has Pfaffian equations vanishing on the graphs of the corresponding Poincaré maps.

A C^r-smooth conjugacy between two local families v and w of the above form is a map

$$H\colon (\mathbb{R}^2,0) \times (\mathbb{R}^k,0) \to (\mathbb{R}^2,0), \quad (x,y,\varepsilon) \mapsto H(x,y,\varepsilon),$$

such that

$$H_* v(x,y,\varepsilon) = w(H(x,y,\varepsilon),\varepsilon),$$

where H_* stands for the Jacobian matrix with respect to the variables x, y. (this definition does not yet allow for reparametrization of a local family). Two families are finitely differentiably equivalent if for any $r < \infty$ there exists a C^r-conjugacy between them. The two families v, w are orbitally equivalent if there exists the

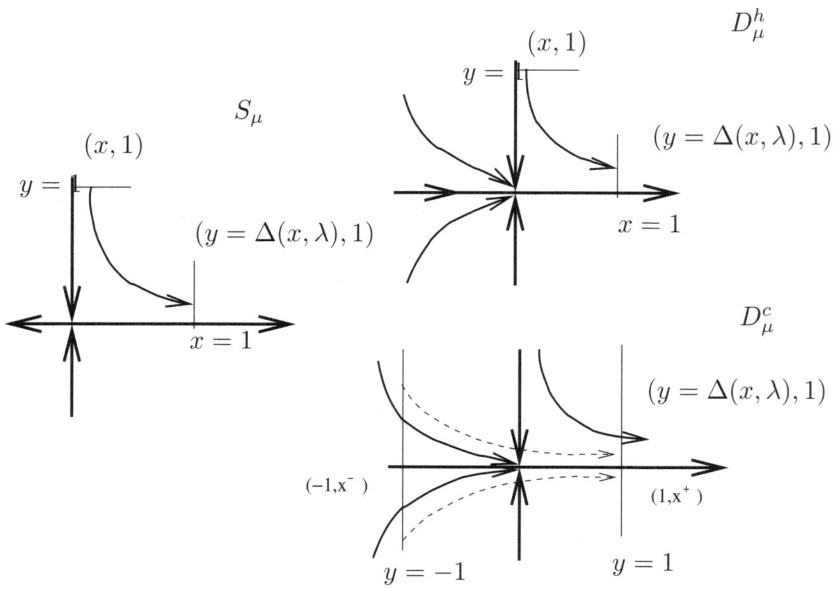

FIGURE 5. Poincaré correspondence maps

germ of a nonvanishing function ϕ: $(\mathbb{R}^2, 0) \times (\mathbb{R}^k, 0) \to \mathbb{R}^1$ such that v is equivalent to $\phi \circ w$.

To allow for a reparametrization of local families, we say that a family $v(\cdot, \varepsilon)$ is induced from another family $w(\cdot, \lambda)$, $\lambda \in (\mathbb{R}^m, 0)$, if $v(\cdot, \varepsilon) = w\big(\cdot, \lambda(\varepsilon)\big)$, where $\lambda(\varepsilon)$ is the germ of a smooth map $(\mathbb{R}^k, 0) \to (\mathbb{R}^m, 0)$. The number of new parameters m may be different from k.

Assume that the family $w(\cdot, \lambda)$ is *global* (i.e., the expression $w(x, y, \lambda)$ makes sense for all $(x, y, \lambda) \in \mathbb{R}^{m+2}$); this happens in particular when w is *polynomial* in all its arguments. Restricting the parameters λ onto a small neighborhood of a certain point $(0, 0, c) \in \mathbb{R}^2 \times \mathbb{R}^m$, we obtain a *localization* of the global family w, which formally becomes a local family after the parallel translation $\lambda \mapsto \lambda - c$.

DEFINITION 12. (1) A local family $v = v(\cdot, \lambda)$ is *finitely smooth orbital versal unfolding* (in short, versal unfolding) of the germ $v(\cdot, 0)$ if any other local family unfolding this germ is finitely differentiable orbitally equivalent to a family induced from v.

(2) A polynomial family $w(\cdot, \lambda)$, $\lambda \in \mathbb{R}^m$, is a *global finitely smooth orbital versal unfolding* (in short, global versal unfolding) for a certain class of local families of vector fields if any local family from this class is finitely differentiable orbitally equivalent to a local family induced from some localization of w.

To investigate a versal unfolding means to investigate at the same time all smooth local finite-parameter families which unfold the same germ $v(\cdot, 0)$. The main result describing versal unfoldings of germs of elementary singularities on the plane is given by the following

THEOREM 4 ([**IY1**]). *Suppose that a generic finite-parameter family of smooth vector fields on the plane possesses an elementary singular point for a certain value* *of the parameters. If this point has at least one hyperbolic sector, than the family is finitely differentiable orbitally equivalent to a family induced from some localization of one of the families given in the second column of Table* 1.

2.1.3. *Singular-regular systems determining the number of limit cycles.* Recall that for simplicity we consider a 2-parameter family of vector fields $\{\dot{x} = v(x,\varepsilon)\}_{\varepsilon\in B^2}$ and suppose that for $\varepsilon = 0$ the vector field $\dot{x} = v(x,0)$ has a polycycle γ which consists of two saddles p_1, p_2 and two separatrices connecting them, γ_1, γ_2 (see Figure 4). For each saddle $\{p_j\}_{j=1,2}$ there is a neighborhood $\{U_j\}_{j=1,2}$ with a C^r*-normal coordinate charts*. Consider transversal segments "entrance" Σ_j^- and "exit" Σ_j^+ which are parallel to coordinate axis of the normal chart such that the phase curve γ_{j-1} enters the neighborhood U_j through Σ_j^- and the phase curve γ_j exists U_j through Σ_j^+. The normal coordinates induce coordinates x_j and y_j on Σ_j^- and Σ_j^+ respectively. For some parameter values the corresponding vector field defines the following collection of Poincaré correspondence maps:

$$
\begin{aligned}
&\Delta_j(\cdot,\varepsilon)\colon x_j \to y_j = \Delta_j(x_j,\varepsilon), && j = 1,2\\
&f_j(\cdot,\varepsilon)\colon y_j \to x_{j+1} = f_j(y_j,\varepsilon), && j = 1,2 \pmod 2,
\end{aligned}
\tag{20}
$$

where $\Delta_j(\cdot,\varepsilon)$ is a local Poincaré map form the "entrance" segment Σ_j^- to the "exit" segment Σ_j^+ and $f_j(\cdot,\varepsilon)$ is a semilocal Poincaré map along the phase curve γ_j form the "exit" segment Σ_j^+ to the "entrance" segment Σ_{j+1}^-.

Now we decompose the monodromy map (the Poincaré first return map) along the polycycle γ into the chain of the two local singular maps $\{\Delta_j(\cdot,\varepsilon)\}_{j=1,2}$ and the semilocal regular maps $\{f_j(\cdot,\varepsilon)\}_{j=1,2}$ of total length 4. Limit cycles correspond to the fixed points of the monodromy. But instead of writing one equation for the fixed points of the monodromy we consider a system of 4 equations, which we call the *preliminary basic system*:

$$
\begin{cases}
y_1 = \Delta_1(x_1,\varepsilon),\\
x_2 - f_1(y_1,\varepsilon) = 0,\\
y_2 = \Delta_2(x_2,\varepsilon),\\
x_1 - f_2(y_2,\varepsilon) = 0.
\end{cases}
\tag{21}
$$

Recall that x_j's are C^r-normal coordinates on Σ_j^- and y_j's are C^r-normal coordinates on Σ_j^+. Thus the system involves C^r-smooth regular functions f_j's and the maps Δ_j from the list (modulo reparametrization $\varepsilon \to \lambda(\varepsilon)$), that are essentially singular. The problem now is to estimate the number of small isolated solutions uniformly over all sufficiently small parameter values.

Suppose for $\varepsilon = \varepsilon^*$ the system (21) has the maximal number of isolated solutions. Since each isolated solution of this system corresponds to an isolated solution of the 1-dimensional Poincaré return map $\Delta(x_1,\varepsilon) = x_1$, one can choose a small δ_1 so that the number of regular (nondegenerate) solutions of $\Delta(x_1,\varepsilon) = x_1 + \delta_1$ bounds the number of isolated solutions to (21) from above (see Figure 6, cf. [**IY2**, Figure 8].)

Recall that a point $x \in \mathbb{R}$ is *nondegenerate* or *regular* for the map Δ if the derivative $\Delta'(x) \neq 0$ in the 1-dimensional case and $x \in \mathbb{R}^n$ is a regular point of a smooth map $F\colon \mathbb{R}^n \to \mathbb{R}^m$ if the rank of the linearization $dF(x)$ at x is maximal. Direct calculation shows that a regular solution to $\Delta(x_1,\varepsilon) = x_1$ (resp.

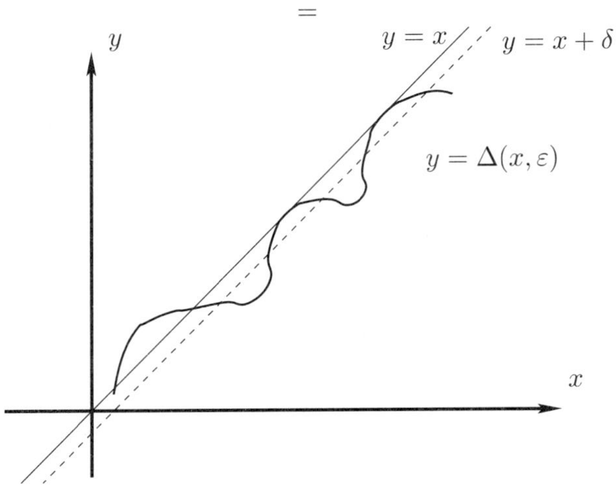

FIGURE 6. Isolated and regular solutions

$\Delta(x_1,\varepsilon) = x_1 + \delta_1$) corresponds to a regular solution to the system (21) (see [**IY2**, Lemma 3.3]).

Moreover, if δ_2 is nonzero and much smaller than δ_1, then by the implicit function theorem the number of regular solutions of the system

$$\begin{cases} y_1 = \Delta_1(x_1,\varepsilon), \\ x_2 - f_1(y_1,\varepsilon) = \delta_1, \\ y_2 = \Delta_2(x_2,\varepsilon), \\ x_1 - f_2(y_2,\varepsilon) = \delta_2 \end{cases} \tag{22}$$

is the same as that for $\Delta(x_1,\varepsilon) = x_1 + \delta_1$. Therefore, it suffices to estimate the number of small regular solutions to the system (22) provided that $1 \gg |\delta_1| \gg |\delta_2| \geq 0$.

2.2. The reduction method of Khovanskiĭ.

2.2.1. *A mixed singular-regular functional system.* The system (22) is not easy to analyze, because it has the singular functions Δ_j. The first key idea of the second step of the Main Result [**MR, IY2**] is to replace these singular equations in (22) by the singular functional-Pfaffian equations which have polynomial differentials of the form (18). As a result we can obtain the *mixed* functional singular-regular system of the following form

$$\begin{cases} \mathcal{F}_1(x_1,y_1,\varepsilon) = 0 \\ F_1(x,y,\varepsilon) = \delta_1, \\ \mathcal{F}_2(x_1,y_1,\varepsilon) = 0, \\ F_2(x,y,\varepsilon) = \delta_2, \end{cases} \tag{23}$$

$$d\mathcal{F}_j(x_j,y_j,\varepsilon) = P_j(x_j,y_j,\varepsilon)\,dx_j + Q_j(x_j,y_j,\varepsilon)\,dy_j,$$

$$F_j(x,y,\varepsilon) = x_{j+1} - f_j(y_j,\varepsilon), \quad X = (x_1,y_1,x_2,y_2) \in (\mathbb{R}^4,0), \quad \varepsilon \in (\mathbb{R}^2,0),$$

where $\mathcal{F}_j$ are functions such that their differentials are polynomial 1-forms of the type from column 4 of Table 1 and $1 \gg |\delta_1| \gg |\delta_2| \geq 0$. In order to simplify considerations we replaced the eigenfunctions $\lambda_j(\varepsilon)$ (see (19)) by ε. What we are interested in is the upper estimate for the number of small regular solutions to (23), uniform over all parameters and all sufficiently small values of δ's with $1 \gg |\delta_1| \gg |\delta_2| \geq 0$.

2.2.2. *Reduction of the mixed functional system* (23) *to chain maps* $P \circ F$. Let $F = (F_1, F_2)\colon \mathbb{R}^4 \to \mathbb{R}^2$ be a smooth map formed by functions F_1 and F_2. Denote by $J^m(\mathbb{R}^4, \mathbb{R}^2)$ the space of k-jets of maps of $\mathbb{R}^4$ to $\mathbb{R}^2$. Fix coordinates in the source $X = (x_1, \dots, x_4)$ and the target (δ_1, δ_2). Then the space $J^m(\mathbb{R}^4, \mathbb{R}^2)$ consists of coordinates in the source, the target, and all partial derivatives of F of order at most k

$$(24) \quad \Bigg\{(x_1, \dots, x_4); \big(F_1(X), F_2(X)\big); \\ \left(\frac{\partial^\alpha F_i}{\partial^{\alpha_1} x_1 \cdots \partial^{\alpha_4} x_4}, \forall i = 1, 2, \alpha_j \geq 0, \text{ such that } \sum_{j=1}^{4} \alpha_j \leq m\right)\Bigg\}.$$

We shall call these coordinates on the m-jet space $J^m(\mathbb{R}^4, \mathbb{R}^2)$ the *natural coordinates.* With this coordinate system the space of m-jets has a natural linear structure. We also denote by $j^m F$ the mth jet of the map F. Denote also by $B_r(0) \subset \mathbb{R}^4$ the r-ball centered at the origin. We call a polynomial map $P\colon \mathbb{R}^N \to \mathbb{R}^n$ nontrivial if the image $P(\mathbb{R}^N)$ has nonempty interior in $\mathbb{R}^n$.

Our goal now is to realize Step 2 of our program outlined in Section 1.3, i.e., estimate the small number of solutions to (23) via geometric multiplicity of the chain maps (12) or prove the following

THEOREM 5 (cf. [**K6**, Theorem 10)]). *Suppose that the degrees of polynomial 1-forms from* (23) *are bounded by some $d \in \mathbb{Z}_+$. Then for a sufficiently small $r > 0$ there exists a set of* 3 (= *the number of singular equations*+1) *explicitly computable nontrivial polynomials $P^k = (P^k_1, \dots, P^k_2)\colon J^2(\mathbb{R}^4, \mathbb{R}^2) \to \mathbb{R}^2$, $k = 0, 1, 2$ defined on the space of* 2*-jets $J^2(\mathbb{R}^4, \mathbb{R}^2)$ such that for a generic C^3 smooth*[3] *map $F\colon \mathbb{R}^4 \to \mathbb{R}^2$ the number of regular solutions to the system* (23) *inside the ball $B_r(0)$ is bounded by the number of small regular solutions*

$$(25) \quad \#\{X \in B_r(0) : (F_1, F_2)(X) = (\delta_1, \delta_2), (P^0_1, P^0_2) \circ j^2 F(X) = (\delta_3, \delta_4)\} \\ + \frac{1}{2} \sum_{k=1,2} \#\{X \in B_r(0) : (F_1, F_2)(X) = (\delta_1, \delta_2), (P^k_1, P^k_2) \circ j^2 F(X) = (\delta_3, \delta_4)\},$$

where $1 \gg |\delta_1| \gg \cdots \gg |\delta_4| \geq 0$ decrease to zero sufficiently fast. The degrees of the polynomials satisfy the inequalities $\deg P^k_i \leq 2^i(d+1)$ for all k and i.

REMARK 2. We cannot find a direct reference in the book of Khovanskiĭ [**K2**], but this theorem is in the spirit of the results about perturbations discussed in Section 5.2 of this book. In fact this theorem is due to Khovanskiĭ.

2.2.3. *An application of Khovanskiĭ's method to the system* (23) *or a proof of Theorem* 5. The method is based on the following version of Rolle's lemma

LEMMA 1. *Consider C^2 functions $f\colon S^1 \to \mathbb{R}^1$ on the circle and $g\colon [0, 1] \to \mathbb{R}^1$ on the segment, with a finite number of critical points. Then for any $a \in \mathbb{R}$ and*

[3]Required smoothness 3 = the number of singular equations + 1

any sufficiently small $\delta > 0$

$$(26)\qquad \begin{aligned} \#\{x : f(x) = a\} &\leq \#\{x : f'(x) = \delta\} \\ \#\{x : g(x) = a\} &\leq \#\{x : g'(x) = \delta\} + 1. \end{aligned}$$

PROOF. One proves first the formula for $\delta = 0$ using the fact that between any two consecutive preimages there is a point where the derivative is zero. Then one uses the nondegeneracy of critical points. □

Now using this lemma we shall prove Theorem 5.

PROOF OF THEOREM 5. Denote by $\rho_r(x) = r - \sum_j x_j^2$ the function which measures distance to the boundary of the r-ball $B_r(0)$ and vanishes on the boundary $\partial B_r(0)$. Recall that r is sufficiently small.

Denote by $G_1 : \mathbb{R}^4 \to \mathbb{R}^3$ the map defined by the coordinate functions $(\mathcal{F}_2, F_1, F_2)$. Then the system (23) under investigation becomes the map $(\mathcal{F}_1, G_1)\colon \mathbb{R}^4 \to \mathbb{R}^4$, given by its coordinate functions. In terms of this map we need to estimate the number of small preimages of points of the form

$$\#\{(\mathcal{F}_1, G_1)^{-1}(0, 0, \delta_1, \delta_2) \cap B_r(0)\},$$

where $1 \gg |\delta_1| \gg |\delta_2| \geq 0$.

Let's estimate the number of small preimages of a point

$$(27)\qquad \#\{(\mathcal{F}_1, G_1)^{-1}(a_1, a_2, \delta_1, \delta_2) \cap B_r(0)\},$$

where $1 \gg |\delta_1| \gg |\delta_2| \geq 0$ and a_1, a_2 are arbitrary. Since there is no restriction on a_1, a_2 the number of solutions may only increase.

Step 1. Eliminate one singular equation, say, $\mathcal{F}_1 = 0$ and replace it by two chain-type equations $\{P_1^i \circ j^1 F\}_{i=0,1}$ so that for a sufficiently small $|\delta_2| \gg |\delta_3| \geq 0$ the number of small regular preimages

$$(28)\qquad \#\big\{\big(G_1, P_1^i \circ j^1 F(X)\big)^{-1}(a_2, \delta_1, \delta_2, \delta_3) \cap B_r(0)\big\}_{i=0,1}$$

is at least the number of small regular preimages of (27) for any a_1, i.e.,

$$(29)\qquad \begin{aligned} &\#\{X \in B_r(0) : (\mathcal{F}_1, G_1)(X) = (a_1, a_2, \delta_1, \delta_2)\} \\ &\qquad \leq \#\{X \in B_r(0) : (G_1, P_1^0 \circ j^1 F)(X) = (a_2, \delta_1, \delta_2, \delta_3)\} \\ &\qquad\qquad + \tfrac{1}{2}\#\{X \in B_r(0) : (G_1, \rho_r)(X) = (a_2, \delta_1, \delta_2, \delta_3)\} \end{aligned}$$

Consider a regular value $(a_2, \delta_1, \delta_2) \in \mathbb{R}^3$ of the map G_1. By the rank theorem [**GG**] the level set $L_{(a_2,\delta_1,\delta_2)} = G_1^{-1}(a_2, \delta_1, \delta_2) \cap B_r(0) \subset B_r(0)$ is a smooth 1-dimensional manifold in the r-ball. It consists of a finite number of connected parts either compact — topological circles, denoted by $\{S_i\}_{i \in I(a_2,\delta_1,\delta_2)}$, or noncompact — curves $\{L_j\}_{j \in J(a_2,\delta_1,\delta_2)}$ reaching the boundary $\partial B_r(0)$. It is easy to see that

$$(30)\qquad \begin{aligned} &\#\{X \in B_r(0) : (\mathcal{F}_1, G_1)(X) = (a_1, a_2, \delta_1, \delta_2)\} \\ &\quad = \sum_{i \in I(a_2,\delta_1,\delta_2)} \#\{X \in S_i : \mathcal{F}_1(X) = a_1\} + \sum_{j \in J(a_2,\delta_1,\delta_2)} \#\{X \in L_j : \mathcal{F}_1(X) = a_1\}. \end{aligned}$$

Let us estimate the first sum on the right-hand side. Fix a circle, say, S_1. Restrict the function $\mathcal{F}_1$ to S_1 and denote the result by $f_1 = \mathcal{F}_1|_{S_1} \colon S^1 \to \mathbb{R}$ (see Figure 7). We get a function f_1 on the circle. Notice that the condition $f_1'(X) = 0$

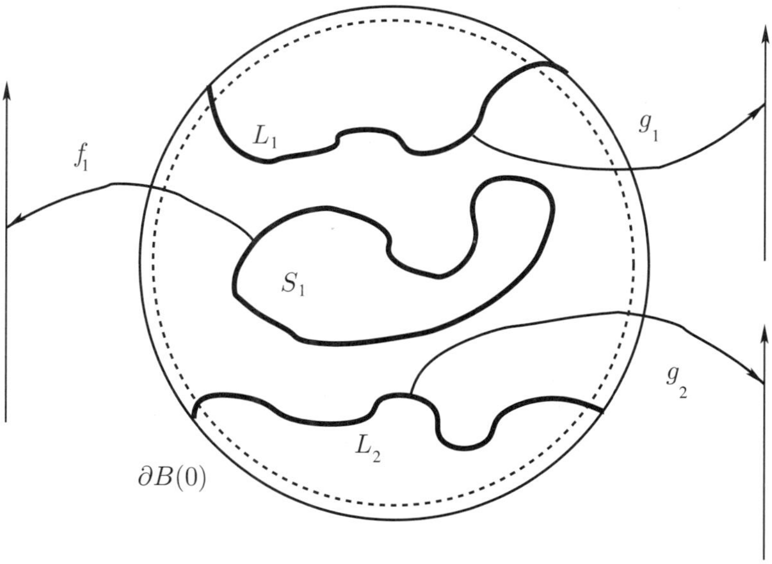

FIGURE 7. Application of Rolle's lemma

is equivalent to the condition that the Jacobian of the map $(\mathcal{F}_1, G_1)$, denoted by $J_{\mathcal{F}_1,G_1}(X)$, is zero.

$$\boxed{f_1'(X) = 0} \iff \boxed{J_{\mathcal{F}_1,G_1}(X) = 0} \tag{31}$$

Recall now that the differentials $d\mathcal{F}_j(X) = P_j\,dx_j + Q_j\,dy_j$, $j = 1, 2$ are polynomial, therefore, we have

$$\begin{aligned} J_{\mathcal{F}_1,G_1}(X) &= *\big(d\mathcal{F}_1(X) \wedge d\mathcal{F}_2(X) \wedge dF_1(X) \wedge dF_2(X)\big) \\ &= \det\big(\nabla\mathcal{F}_1(X), \nabla\mathcal{F}_2(X), \nabla F_1(X), \nabla F_2(X)\big) = P_1^0 \circ j^1F(X). \end{aligned} \tag{32}$$

where $*$ is a natural isomorphism between the space of functions on $\mathbb{R}^4$ and 4-forms and $\nabla F(X)$ is the gradient vector of a function $F\colon \mathbb{R}^4 \to \mathbb{R}$. Since $\deg P_j, \deg Q_j \leq d$, the degree of P_1^0 is bounded by $2(d+1)$.

Now we can apply Rolle's Lemma 1 with $f = f_1$ and get that for any a_1 and a sufficiently small $\delta_3 \neq 0$

$$\sum_i \#\{X \in S_i : \mathcal{F}_1(X) = a_1\} \leq \sum_i \#\{X \in S_i : J_{\mathcal{F}_1,G_1}(X) = \delta_3\}. \tag{33}$$

The second sum can be estimated in almost the same way. Instead of using Lemma 1 with $f = f_i$ we need to use Lemma 1 with $g = g_j = \mathcal{F}_1|_{L_j}\colon [0,1] \to \mathbb{R}$ (see Figure 7). Denote the number of components reaching the boundary $|J|$ by k. Then

$$\sum_{j=1}^{k} \#\{X \in L_j : \mathcal{F}_1(X) = a_1\} \leq \sum_{j=1}^{k} \#\{X \in L_j : J_{\mathcal{F}_1,G_1}(X) = \delta_3\} + k \tag{34}$$

In order to find the number of components reaching the boundary notice that each such component intersects the sphere $\rho_r^{-1}(\delta_3)$ for $\delta_3 > 0$ in at least two points. So $P_1^1 \circ j^1F(X) = \rho_r(X)$ and the second term in inequality (29) corresponds to the number of noncompact components (the boundary term). This completes the proof of Step 1 or proves (29).

For $i = 0, 1$ denote by $G_2^i\colon \mathbb{R}^4 \to \mathbb{R}^3$ the maps defined by their coordinate functions $(F_1, F_2, P_1^i \circ j^1 F)$. Let's fix $i = 0$ or 1.

Step 2. Eliminate the second singular equation $\mathcal{F}_2 = 0$ and replace it by two chain-type equations $\{P_2^i \circ j^2 F\}_{i=0,1,2}$ so that for a sufficiently small $0 \leq |\delta_4| \ll |\delta_3|$ the number of small regular preimages $\{(\mathcal{F}_2, G_2^i)^{-1}(a_2, \delta_1, \delta_2, \delta_3)\}_{i=0,1}$ is at least the number of small regular preimages $(G_2^i, P_2^i \circ j^2 F)^{-1}(\delta_1, \delta_2, \delta_3, \delta_4)$ for any a_2, i.e.,

$$\begin{aligned}(35)\quad \#\{X \in B_r(0) : (\mathcal{F}_2, G_2^i)(X) &= (a_2, \delta_1, \delta_2, \delta_3)\} \\ \leq \#\{X \in B_r(0) &: (G_2^i, P_2^i \circ j^2 F)(X) = (\delta_1, \delta_2, \delta_3, \delta_4)\} \\ &+ \tfrac{1}{2}\#\{X \in B_r(0) : (G_1, \rho_r)(X) = (\delta_1, \delta_2, \delta_3, \delta_4)\}\end{aligned}$$

The proof of this inequality is very similar to the proof of Step 1. We reproduce a shortened version of it in order to show why *the condition* $|\delta_4| \ll |\delta_3|$ *is necessary* for the Khovanskiĭ method to work.

Let's choose a regular value $(\delta_1, \delta_2, \delta_3)$ for the map G_2^i and consider the level set $L_{(\delta_1,\delta_2,\delta_3)} = \left(G_2^i\right)^{-1}(\delta_1, \delta_2, \delta_3)$ which by the rank theorem is a smooth 1-dimensional manifold consisting of a finite number of connected components either compact — topological circles, denoted by $\{S_i\}_{i \in I(\delta_1,\delta_2,\delta_3)}$, or noncompact — curves $\{L_j\}_{j \in J(\delta_1,\delta_2,\delta_3)}$ reaching the boundary $\partial B_r(0)$.

Then we restrict $\mathcal{F}_2$ to $L_{(\delta_1,\delta_2,\delta_3)}$ and get a finite collection of functions $\{f_i = \mathcal{F}_1|_{S_i} : S^1 \to \mathbb{R}\}_{i \in I}$ on circles and $\{g_j = \mathcal{F}_1|_{L_j} : [0,1] \to \mathbb{R}\}_{j \in J}$ on the interval $[0,1]$. In order to use Rolle's Lemma 1 we need to compute the condition $f_i'(X) = 0$ (resp. $g_j'(X) = 0$). This is equivalent to the Jacobian $J_{\mathcal{F}_2, G_2^i}(X)$ of the map $(\mathcal{F}_2, G_2^i)$ being equal to 0

$$\begin{aligned}(36)\qquad J_{\mathcal{F}_2,G_2^i}(X) &= *\big(d\mathcal{F}_2(X) \wedge dF_1(X) \wedge dF_2(X) \wedge d(P_1^i \circ j^1 F)(X)\big) \\ &= \det\big(\nabla\mathcal{F}_2(X), \nabla F_1(X), \nabla F_2(X), \nabla(P_1^i \circ j^1 F)(X)\big) \\ &= P_2^i \circ j^2 F(X).\end{aligned}$$

Since $d\mathcal{F}_2(X) = P_2\, dx_2 + Q_2\, dy_2$ and $\deg P_2, \deg Q_2 \leq s$, the degree of P_2^i is bounded by $4(s+1)$. An easy calculation shows that each time we take a Jacobian of a chain-map $P \circ j^k F$ its degree at most doubles.

Now we would like to apply Rolle's Lemma 1 with $f = f_i$ (resp. $g = g_j$) and substitute a singular equation $\mathcal{F}_2$ by the equation $J_{\mathcal{F}_2,G_2^i}(X) = \delta_4$. This equation has to be equivalent to the fact that the derivative $f'(X)$ (resp. $g'(X)$) is small or the covectors $\nabla\mathcal{F}_2(X)$, $\nabla F_1(X)$, $\nabla F_2(X)$, and $\nabla(P_1^i \circ j^1 F)(X)$ have to be almost linearly dependent. However, the determinant (36) can be almost zero *not because gradient vectors are almost linearly dependent, but because one of the gradient vectors is small.* In order to avoid this problem let's make the following remark: for a fixed regular value $(\delta_1, \delta_2, \delta_3)$ the level set $L_{(\delta_1,\delta_2,\delta_3)}$ is a smooth compact 1-dimensional manifold possibly with boundary and the lengths of the gradient vectors $\nabla F_1(X), \nabla F_2(X)$, and $\nabla(P_1^i \circ j^1 F)(X)$ have to be bounded away from zero. Knowing how far these lengths are from zero we can choose δ_4 of much smaller size to guarantee almost linear dependence of the gradient vectors. This proves inequality (35).

This argument allows one to apply Rolle's Lemma 1 in the described fashion inductively in any dimension and eliminate an arbitrary number of singular Pfaffian equations. This completes the proof of Theorem 5. See [**IK, K6**] for a more general treatment. □

2.2.4. *Geometric multiplicity of chain maps.* Let $P\colon \mathbb{R}^N \to \mathbb{R}^n$ be a nontrivial vector-polynomial, i.e., the image $P(\mathbb{R}^N) \subset \mathbb{R}^n$ has nonempty interior, let $B^n \subset \mathbb{R}^n$ be a unit ball, and $F\colon B^n \to \mathbb{R}^N$ be a generic sufficiently smooth map with $N \geq n$. We call the composition of a generic smooth map and a nontrivial polynomial

$$P \circ F\colon B^n \to \mathbb{R}^n \tag{37}$$

a *chain map*. More generally, let $P\colon J^n(B^n, \mathbb{R}^N) \to \mathbb{R}^n$ be a nontrivial vector-polynomial, defined on the space of m-jets for some $m \in \mathbb{Z}_+$. Then a *chain map* is

$$P \circ j^m F\colon B^n \to \mathbb{R}^n \tag{38}$$

As a result of the application of Theorem 5 to the system (23) we need to estimate the number of small regular preimages of a special point of a *chain map* or its geometric multiplicity, defined in (8). Actually, application of Theorem 5 gives not a generic smooth map, but a jet of a generic smooth map. To simplify the discussion we consider the case of a smooth map (37). The general jet case can be treated using the same method.

The next two lectures are devoted to a proof of Bézout's theorem for chain maps. Recall that $B_r(0) \subset \mathbb{R}^n$ denotes the r-ball centered at the origin.

THEOREM 6 (cf. [**K6**, Theorem 3]). *Let $P = (P_1, \dots, P_n)$ be a nontrivial polynomial defined on the space of m-jets $P\colon J^m(B^n, \mathbb{R}^N) \to \mathbb{R}^n$ and let $F\colon B^n \to \mathbb{R}^N$ be a C^k smooth mapping, $k > m$, and $N > n$. Suppose F satisfies a transversality condition depending only on P. Then for a sufficiently small r, to find a geometric multiplicity of the chain map (37) at 0 one can replace $j^m F$ by $L_{F,0,m}$, its linear part at 0. Namely,*

$$\begin{aligned} \#\{X \in B_r(0) : P_1 \circ j^m F(X) = \delta_1, \dots, P_n \circ j^m F(X) = \delta_n\} \\ = \#\{X \in B_r(0) : P_1 \circ L_{F,0,m}(X) = \delta_1, \dots, P_n \circ L_{F,0,m}(X) = \delta_n\}, \end{aligned} \tag{39}$$

where $1 \gg |\delta_1| \gg \cdots \gg |\delta_n| \geq 0$.

REMARK 3. By Bézout's theorem the number of isolated solutions to the equation in the right-hand side of (39) can be bounded by the product $\prod_{i=1}^n \deg P_i$.

The classical transversality theorem [**AGV**] says that for any given transversality condition, a generic mapping $F\colon B^n \to \mathbb{R}^N$ satisfies this condition and generic mappings form an open dense set in the space of smooth mapping of B^n to $\mathbb{R}^N$. Moreover, a mapping F with "probability one" satisfies a previously given transversality condition. For definitions of "probability one" or "prevalence" see [**HSY, K1**].

Acknowledgments. I would like to thank Askold Khovanskiĭ whose deep insight helped me to make a significant simplification of the application of Khovanskiĭ's method.

3. Stratifications and Bézout's theorem for chain maps

In this lecture, first in Section 3.1 we describe a geometric picture behind Bézout's theorem for chain maps (Theorem 6) formulated at the end of the last lecture. It turns out that the proof of this theorem reduces to a question about the existence of a certain, so-called, a_P-stratification for the outer part P of the chain map (38). Then in Section 3.2 we define the necessary notions from stratification theory, including a_P-stratification and discuss the question of existence of an a_P-stratification. In general, it does not always exist, as examples from Section 3.3.1 of Thom and Grinberg show. At the end of this section we state Hironaka's theorem on the existence of a_P-stratification for polynomial functions, i.e., for polynomial maps with 1-dimensional image, and its extension, a theorem on the existence of an a_P-stratification for maps with a multidimensional image proven in [**K6**]. Such a theorem is required for the proof of the Main Result. Finally, in Section 3.4 we present a geometric proof of Hironaka's theorem based on the author's proof of the existence of Whitney's stratification [**K7**]. A proof of the existence of Whitney's stratification is also outlined.

3.1. A heuristic description. Consider a chain map $P \circ F\colon \mathbb{R}^2 \to \mathbb{R}^2$, where $F\colon \mathbb{R}^2 \to \mathbb{R}^N$ is a generic C^k smooth map, $N, k > 2$ and $P = (P_1, P_2)\colon \mathbb{R}^N \to \mathbb{R}^2$ is a polynomial of degree d. Fix a small positive r. We would like to estimate the maximal number of small preimages

$$\#\{x \in B_r(0) : P_1 \circ F(x) = \varepsilon, P_2 \circ F(x) = 0\} \tag{40}$$

for a small enough ε.

To show the idea put $N = 3$, $P_1(x,y,z) = x^2 + y^2$, and $P_2(x,y,z) = xy$. Assume also that $F(0) = 0$. Denote the level set by $V_\varepsilon = \{P_1 = \varepsilon, P_2 = 0\}$. The level set V_ε for $\varepsilon > 0$ consists of 4 parallel lines (see Figure 8).

Notice that in our notation the number of intersections of $F\big(B_r(0)\big)$ with V_ε equals the number of preimages of the point $(\varepsilon, 0)$ under $P \circ F$, see (40).

It is easy to see from Figure 8 that if F is transverse to V_0 it is transverse to V_ε for any small $\varepsilon > 0$. Moreover, the number of intersections $F\big(B_r(0)\big)$ with V_ε equals 4 (see the points $P_1, \dots, P_4$).

Another way to calculate the same number is as follows. Let us replace F by its linear part L_F at zero. Then

$$\begin{aligned}\#\{x \in B_r(0) &: P_1 \circ F(x) = \varepsilon, P_2 \circ F(x) = 0\} \\ &= \#\{x \in B_r(0) : P_1 \circ L_F(x) = \varepsilon, P_2 \circ L_F(x) = 0\}\end{aligned}$$

and solving this polynomial system also yields 4.

The idea behind this picture is the following: Consider an arbitrary N and a polynomial $P = (P_1, P_2)\colon \mathbb{R}^N \to \mathbb{R}^2$ of degree at most d and $N > 2$. Define the algebraic variety $V_\varepsilon = (P_1, P_2)^{-1}(\varepsilon, 0)$ as the level set.

Assume for simplicity that for any small $\varepsilon \neq 0$ the level set V_ε is a manifold of codimension 2. We shall get rid of this assumption later (see Theorem 9(b)). It turns out that there exists a special partition $\mathcal{V}_0 = \{V_i\}_{i \in \mathcal{A}}$ of $V_0 = \bigsqcup_{i \in \mathcal{A}} V_i$ into semialgebraic parts which are attached to their neighbors "regularly" see Definition 14) such that it depends only on P and satisfies the following condition. We say F is transverse to a stratified set $(V_0, \mathcal{V}_0)$ if F is transverse to each stratum

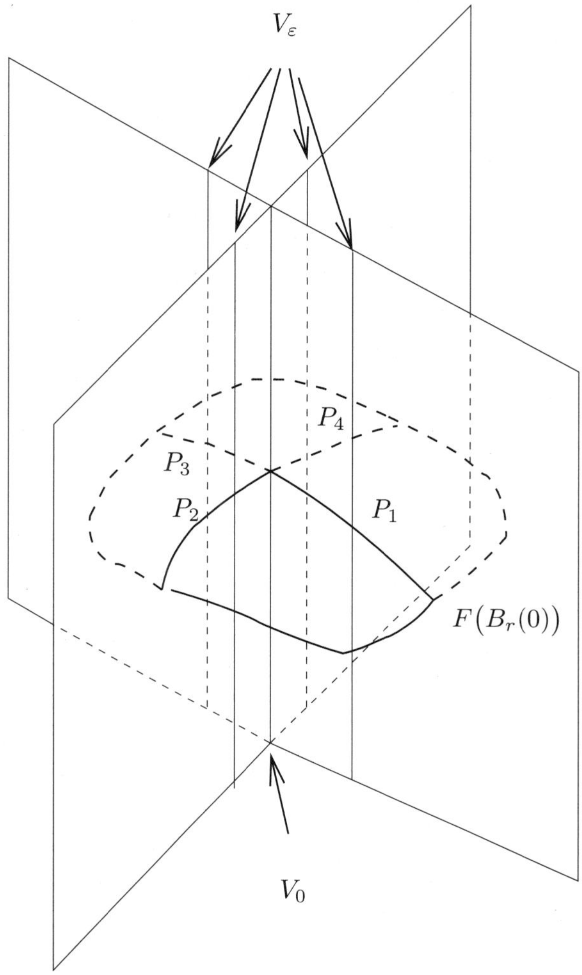

FIGURE 8. The Idealistic Example

$V_i \in \mathcal{V}_0$; then

(41) $$\boxed{F \text{ is transverse to } (V_0, \mathcal{V}_0)} \implies \boxed{F \text{ is transverse to } V_\varepsilon}$$

LEMMA 2. *Let $B_r(0)$ be the r-ball centered at the point $0 \in \mathbb{R}^2$ and let $L_{F,0}$ denote the linearization of F at the point a. Under condition (41), the number of intersections of the image $F\big(B_r(0)\big)$ with V_ε coincides with the number of intersections of the image $L_{F,0}\big(B_r(0)\big)$ with V_ε, provided r is small enough. That is:*

(42) $$\#\{x \in B_r(0) : (P_1, P_2) \circ F(x) = (\varepsilon, 0)\} \\ = \#\{x \in B_r(0) : (P_1, P_2) \circ L_{F,0}(x) = (\varepsilon, 0)\}.$$

REMARK 4. The argument below is independent of the codimension of V_ε. We only need condition (41) and the fact that the codimension of V_ε coincides with the dimension of the preimage of a chain map $P \circ F$.

PROOF. Consider the 1-parameter family of maps $F_t = tF+(1-t)L_F$ deforming the linear part of F into F. Clearly, $F_1 \equiv F$ and $F_0 \equiv L_F$. Fix a small $r > 0$. Since F is transverse to V_0 at 0 all F_t are transverse to V_0 at 0. Condition (41) implies that for all small ε and all $t \in [0,1]$ we have F_t is transverse to V_ε.

Therefore, *the number of intersections of $F_t(B_r(0))$ with V_ε is independent of t.* Indeed, assume that $\#\{F_{t_1}(B_r(0)) \cap V_\varepsilon\} \neq \#\{F_{t_2}(B_r(0)) \cap V_\varepsilon\}$ for some $t_1 < t_2$. Then as t_1 increases to t_2 there is a point t^* where the number of intersections either drops or jumps. At this point t^* the condition of transversality of F_{t^*} and V_ε must fail. This completes the proof of the lemma. □

3.2. Basic definitions of stratified sets, maps, etc.

3.2.1. *Stratified sets.* A stratification of a set, e.g., an analytic variety, is, roughly, a partition of it into manifolds so that these manifolds fit together "regularly." Stratification theory was originated by Thom [**T2**] and Whitney [**W2**] for algebraic and analytic sets. It was one of the key ingredients in Mather's proof of the topological stability theorem [**M2**]. For the history and further applications of stratification theory see [**GM, dPW**].

We consider here only the category of real (semi)algebraic sets for simplicity. Theorems on existence of stratifications proven here in the category of semialgebraic sets can be proven for the categories of complex or real (semi)analytic sets using similar methods. Call a subset $V \subset \mathbb{R}^N$ a *semivariety* if locally at each point $x \in \mathbb{R}^N$ it is a finite union of subsets defined by equations and inequalities

$$f_1 = \cdots = f_k = 0 \quad g_1 > 0, \ldots, g_l > 0 \tag{43}$$

where f_i's and g_j's are real algebraic depending on x. Semivarieties are closed under Boolean operations.

DEFINITION 13 (Whitney). Let V_i, V_j be disjoint manifolds in $\mathbb{R}^N$, $\dim V_j > \dim V_i$, and let $x \in V_i \cap \overline{V_j}$. A triple (V_j, V_i, x) is called a(resp. b)-regular if

(a) when a sequence $\{y_n\} \subset V_j$ tends to x and $T_{y_n}V_j$ tends in the Grassmannian bundle to a subspace τ_x of $\mathbb{R}^N$, then $T_xV_i \subset \tau_x$;
(b) when sequences $\{y_n\} \subset V_j$ and $\{x_n\} \subset V_i$ each tends to x, the unit vector $(x_n-y_n)/|x_n-y_n|$ tends to a vector v, and $T_{y_n}V_j$ tends to τ_x, then $v \in \tau_x$.[4]

V_j is called a(resp. b)-regular over V_i if each triple (V_j, V_i, x) is a(resp. b)-regular.

REMARK 5. Since the Grassmannian manifold of $\dim V_j$-planes in m-dimensional space is compact, existence of limits in the definition above can be reached by choosing a subsequence $\{x_{n_k}\}_{k\in\mathbb{Z}_+}$ or $\{y_{n_k}\}_{k\in\mathbb{Z}_+}$ if necessary.

DEFINITION 14 (Whitney). Let V be a semivariety in $\mathbb{R}^N$. A disjoint decomposition

$$V = \bigsqcup_{i\in I} V_i, \quad V_i \cup V_j = \varnothing \text{ for } i \neq j \tag{44}$$

into smooth semivarieties $\mathcal{V} = \{V_i\}_{i\in I}$, called strata, is called an a(resp. b)-regular stratification if

(1) each point has a neighborhood intersecting only finitely many strata;

---

[4]This way of defining b-regularity is due to Mather [**M2**]. Whitney's definition [**W2**] is equivalent to this one provided a-regularity holds.

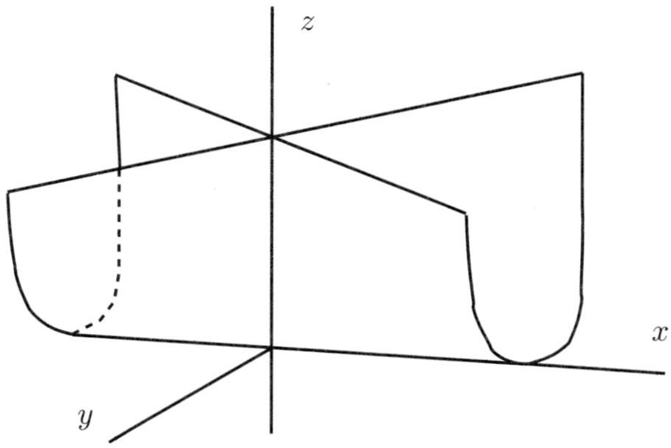

FIGURE 9. Whitney's umbrella

(2) the frontier $\overline{V_j} \setminus V_j$ of each stratum V_j is a union of other strata $\bigsqcup_{i \in J(i)} V_i$;
(3) any triple (V_j, V_i, x) such that $x \in V_i \subset \overline{V_j}$ is a(resp. b)-regular.

The classical example of a stratified algebraic set in $\mathbb{R}^3$ is the so-called *Whitney umbrella.* It is defined as follows

EXAMPLE 2. Consider the 2-dimensional algebraic variety in $\mathbb{R}^3$, defined by

$$V = \{(x, y, z) \in \mathbb{R}^3 : y^2 = zx^2\}. \tag{45}$$

The first natural partition of V into smooth parts (strata) is the vertical line $V_1 = \{x = y = 0\}$ and the complement $V_2 = V \setminus V_1$. However, V_2 does not fit regularly to V_1 at the origin. To see that consider the sequence of the form $(x_n, 0, 0) \in V_2$ with $x_n \to 0$ as $n \to \infty$. It is easy to see that after we refine V_1 into $V_{0'} = \{0\}$ and $V_{1'} = V_1 \setminus V_0$ and put $V_2 = V_{2'}$ the partition $V = \bigsqcup_{i'=1,2,3} V_{i'}$ becomes the stratified manifold $(V, \{V_{0'}, V_{1'}, V_{2'}\})$.

THEOREM 7 ([**W2, T3, Ł1**]). *For any semivariety V in $\mathbb{R}^N$ there is an a(resp. b)-regular stratification.*

REMARK 6. This theorem is not true for smooth sets. To see that one can construct a 2-surface in $\mathbb{R}^3$ which looks like a corkscrew.

Existence of stratifications in the complex analytic case was proved by Whitney [**W2**]. Later Thom published a sketch of a proof [**T3**]. Then Łojasiewicz [**Ł1**] extended these results to the semianalytic case. Later other proofs were found. In [**H**] Hironaka found a nice proof using his resolution of singularities. J. Bochnak, M. Coste, and M.-F. Roy [**BCR**] and Z. Denkowska, K. Wachta [**DW**] follow the classical route of Whitney [**W2**] via the wing lemma. [**BCR**] uses a Nash wing lemma and [**DW**] apply the parameterized Puiseux theorem of W. Pawlucki [**P1**]. T. Wall [**W1**] and S. Łojasiewicz, J. Stasica, and K. Wachta [**ŁSW**] found proofs which use Milnor's curve selection lemma [**M3**]. The latter proof also uses the subanalyticity of the tangent map (for which an elementary proof was given by Z. Denkowska and K. Wachta [**DW**]). In [**K7**] the author gives a geometric proof based on the simple observation that regularity of stratifications is related to uniqueness of the limit of the tangent planes to a bigger stratum as they approach a smaller stratum. This proof is outlined in Section 3.4.1. For a nice exposition of the theory of semianalytic and subanalytic sets see [**Ł2**].

3.2.2. *Stratified maps and* a_P*-stratification.* First we define a smooth map of a stratified set $(V, \mathcal{V})$:

DEFINITION 15. Let $(V, \mathcal{V})$ be a stratified set in $\mathbb{R}^N$, $V \subseteq \mathbb{R}^N$; then a map $P \colon V \to \mathbb{R}^n$ is called C^2-smooth if it can be extended to a C^2 smooth map of an open set U, $V \subset U \subset \mathbb{R}^N$, denoted by $\mathbf{P} \colon U \to \mathbb{R}^n$, whose restriction to V coincides with P.

A stratification $V = \bigcup_i V_i$ stratifies a smooth map $P \colon V \to \mathbb{R}^n$ if the restriction of P to a stratum V_i has constant rank or rank $dP|_{V_i}(x)$ is independent of $x \in V_i$.

A map $G \colon \mathbb{R}^n \to \mathbb{R}^N$ is called transverse to a stratified set $(V, \mathcal{V})$ if G is transverse to each stratum $V_i \in \mathcal{V}$.

EXAMPLE 3. With the notations of the Example 2 of the Whitney umbrella consider the Whitney umbrella V and the projection $P = \pi|_V \colon V \to \mathbb{R}^2$ along the z-coordinate restricted to it. Then the stratification $(V, \{V_{0'}, V_{1'}, V_{2'}\})$ of V is also an a_P-stratification of $P \colon V \to \mathbb{R}^2$, a term to be defined below.

Let V_i be a stratum of a stratification $(V, \mathcal{V})$, $\mathcal{V} = \{V_i\}_{i \in \mathcal{A}}$ and $a \in N$. Denote by $L_{a,i} = (P^{-1}(a) \cap V_i)$ the level sets of P in V_i. By the rank theorem [**GG**], if a stratification $(V, \mathcal{V})$ stratifies a smooth map P, then for each stratum V_i the number $d_i(P) = \dim V_i - \operatorname{rank} dP|_{V_i}$ is well defined and equals the dimension of any nonempty level set $L_{a,i}$.

Roughly speaking, an a_P-stratification is a stratification of a map $P \colon V \to N$ such that it is also an a-stratification of its level sets, i.e., for any sequence of points $\{b_k\} \subset P(V_j)$ converging to a point $a \in P(V_i)$ the corresponding level sets $L_{b_k,j} = (P^{-1}(b_k) \cap V_j) \subset V_j$ approach the limiting level set $L_{a,i} \subset V_i$ "regularly." A precise definition is the following

DEFINITION 16. Let $P \colon \mathbb{R}^N \to \mathbb{R}^n$ be a C^2 smooth map and let V_j and V_i be submanifolds of $\mathbb{R}^N$ such that $V_i \subset \overline{V_j} \setminus V_j$ and the restrictions $P|_{V_j}$ to V_j and $P|_{V_i}$ to V_i have constant ranks. V_j is called a_P-regular over V_i with respect to the map P at a point $x \in V_i \cap \overline{V_j}$ if for any sequence $\{x_n\} \subset V_j$ converging to x the sequence of tangent planes to the level sets $T_n = \ker dP|_{V_j}(x_n)$ converges in the corresponding Grassmannian manifold of $\dim \ker dP|_{V_j}$-dimensional planes to a plane τ and

$$\lim \ker dP|_{V_j}(x_n) = \tau \supseteq \ker dP|_{V_i}(x) \tag{46}$$

DEFINITION 17 (Thom). A C^2 smooth map $P \colon V \to \mathbb{R}^n$ of a stratifiable set V to $\mathbb{R}^n$ is called a_P-stratifiable if there exists a stratification $(V, \mathcal{V})$ such that the following conditions hold:

(a) $(V, \mathcal{V})$ stratifies the map P (see Definition 15);
(b) for all pairs V_j and V_i from $\mathcal{V}$ such that $V_i \subseteq \overline{V_j} \setminus V_j$ the stratum V_j is a_P-regular over the stratum V_i with respect to P at the point x for all $x \in V_i \cap \overline{V_j}$.

REMARKS. (1) The original definition of a_P-stratification requires an appropriate stratification of the image too [**M2**], but for simplicity we do not require the stratification of the image and this weaker definition turns out to be sufficient for our purposes.

(2) With the notations above for an a_P-stratification to exist we must have $d_i(P) \le d_j(P)$ for each $V_i \subseteq \overline{V_j} \setminus V_j$, i.e., nonempty level sets $L_{b,j}$ inside the bigger stratum V_j have dimension $d_j(P)$ greater or equal to the dimension $d_i(P)$ of

nonempty level sets $L_{a,i}$ in the smaller stratum V_i. Otherwise $\dim\ker dP|_{V_j}(x_n) < \dim\ker dP|_{V_i}(x)$ and condition (46) can't be satisfied.

3.2.3. *For* a_P*-stratifications condition* (41) *holds.* The heuristic description given above shows that the key to a proof of Bézout's theorem is condition (41) (Lemma 2). Now we prove that existence of an a_P-stratification of a polynomial map P is sufficient for condition (41) to hold.

Let $P = (P_1, P_2)\colon \mathbb{R}^N \to \mathbb{R}^2$ be a nontrivial vector-polynomial, i.e., the image $P(\mathbb{R}^N)$ has nonempty interior. Denote by $V = P_2^{-1}(0)$ and $V_0 = (P_1, P_2)^{-1}(0)$ these level sets. Assume that there exists a stratification $(V, \mathcal{V})$ which stratifies the map $P|_V$ and such that the zero level set V_0 is also stratified by a stratification $(V_0, \mathcal{V}_0)$ with $V_0 = \bigsqcup_{i\in\mathcal{A}_0} V_i$

LEMMA 3. *With the above notation, if each stratum* $V_j \in \mathcal{V}\setminus\mathcal{V}_0$ *is* a_P*-regular over each stratum* $V_i \in \mathcal{V}_0$ *with respect to the polynomial* P*, then any* C^2 *smooth map* $F\colon \mathbb{R}^2 \to \mathbb{R}^N$ *which is transverse to* $(V_0, \mathcal{V}_0)$ *is also transverse to each level set* $\mathcal{V}_{b;j}$ *for any small* b *and this is equivalent to condition* (41).

PROOF. Pick a point x in $V_i \subset V_0$ and a point $y \in V_i$. Notice that $\ker dP|_{V_i}(x)$ is the tangent plane to the level set $\{P^{-1}(P(x))\cap V_i\}$ at the point x and $\ker dP|_{V_j}(y)$ is the tangent plane to the level set $\{P^{-1}(P(y)) \cap V_j\}$.

By condition (46), if a map $F : X \to \mathbb{R}^N$ is transverse to $\ker dP|_{V_i}(x)$ at a point x, then F is transverse to $\ker dP|_{V_j}(y)$ for any $y \in V_j$ nearby x. This completes the proof of the lemma. □

3.3. Existence of a_P-stratifications for polynomial maps.

3.3.1. *Examples of nonexistence due to Thom and Grinberg.* Existence of a_P-stratifications is a nontrivial question. There are some obvious obstacles. For example, let $V \subset \mathbb{R}^N$ be an algebraic variety and let $P\colon \mathbb{R}^N \to \mathbb{R}^n$ be a polynomial map. Assume that $(V, \mathcal{V})$ stratifies P. Take two strata V_i and V_j so that V_j lies "over" V_i, i.e., $V_i \subseteq \overline{V_j}\setminus V_j$; then the condition (46) can't be satisfied if the dimension of the level sets $d_i(P)$ in the lower stratum V_i is strictly less than that of $d_j(P)$ in the lower stratum V_j, i.e., $\dim\ker dP|_{V_i}(y) < \dim\ker dP|_{V_j}$. In this case a plane $\ker dP|_{V_j}(x)$ of the upper stratum V_j should belong to a plane τ of smaller dimension by condition (46), which is impossible. Thom constructed the first example for which this happens [**GWdPL**].

EXAMPLE 4 (Thom). Consider the vector-polynomial P of the form

$$P\colon \begin{pmatrix} x \\ y \end{pmatrix} \to \begin{pmatrix} x \\ xy \end{pmatrix}. \tag{47}$$

The line $\{x = 0\}$ is the line of critical points of P. Outside of the line $\{x = 0\}$ the map P is a diffeomorphism; therefore, the preimage $P^{-1}(a)$ of any point $a \neq 0$ is 0-dimensional. On the other hand, the preimage of 0 is the 1-dimensional line $\{x = 0\}$. Thus, a_P-regularity fails to exist.

DEFINITION 18. Let us call an algebraic set V rank compatible with respect to a polynomial P if there exists a stratification $(V, \mathcal{V})$ which stratifies P and for any pair V_i and V_j from $\mathcal{V}$ such that $V_i \subseteq \overline{V_j}\setminus V_j$, the dimension of the level sets $d_i(P)$ in the lower stratum V_i does not exceed the dimension of the level sets $d_j(P)$ in the upper stratum V_j.

It turns out that even if an algebraic set V is rank compatible with respect to a polynomial P, then an a_P-stratification still does not always exist. Let us present an example with this property due to M. Grinberg. It seems that the existence of a counterexample was known before, but we did not find an appropriate reference.

Let $V = \{(x, y, z, t) \in \mathbb{R}^4 : x^2 = t^2y + z\}$ be the three-dimensional algebraic variety and $P : V \to \mathbb{R}^2$ be the natural projection associated to the last two coordinates, i.e., $P : (x, y, z, t) \to (z, t)$.

LEMMA 4. *With the above notations the set V is rank compatible with the polynomial map P and does not have an a_P-stratification.*

PROOF. Consider a rank stratification of V. Such a stratification consists of three strata: $V_1 = \{x = t = z = 0\}$, $V_2 = \{t = 0, x^2 = z, x \neq 0\}$, and $V_3 = \{t \neq 0\}$. On each stratum $\operatorname{rank} P|_{V_i} = i - 1$. The level sets $P^{-1}(t, z)$ are parabolas for $t \neq 0$ and lines for $t = 0$.

Show that for each point $\mathbf{a} = (0, a, 0, 0) \in V_1$ there exists a 1-parameter family of level sets such that at the point $\mathbf{a}$ the property of a_P-regularity of V_3 over V_1 fails.

Consider the preimage of the curve $\{z = -at^2\} \subset \mathbb{R}^2$. This is an algebraic variety of the form $W_a = \{x^2 = t^2(y - a)\}$. One can see that W_a is the Whitney umbrella (see Figure 9). The level set $x^2 = t_0^2(y - a)$ is a parabola. As $t_0 \to 0$ this parabola tends to the semiline $\{x = t = z = 0, y \geq a\}$. At the point $\mathbf{a} \in V_1$ the property of a_P-regularity of V_3 over V_1 clearly fails. This completes the proof of the lemma. □

3.3.2. *Existence of a_P-stratifications (Hironaka's theorem and its extension).* As we have seen above, sometimes a_P-stratifications exist, sometimes they do not. Let us state a positive result on their existence.

THEOREM 8 (Hironaka [**H**]). *If $V \subset \mathbb{R}^N$ is a semialgebraic variety and $P \colon \mathbb{R}^N \to \mathbb{R}$ is a polynomial function, then there exists an a_P-stratification $(V, \mathcal{V})$ of V with respect to P by semialgebraic strata.*

In the next section we give a geometric proof of this result based on the proof of the existence of Whitney's stratifications due to the author [**K7**]. Below we describe an extension of Hironaka's theorem to maps with a multidimensional image proven in [**K6**]. This extension is sufficient to prove Bézout's theorem for chain maps (Theorem 6).

TARSKI – SEIDENBERG PRINCIPLE (see, e.g., [**BCR, J**]). *For any semialgebraic V in $\mathbb{R}^N$ and a polynomial map $P \colon \mathbb{R}^N \to \mathbb{R}^n$ the image $P(V)$ is semialgebraic.*

Let $\mathbb{R}^N$ and $\mathbb{R}^n$ be Euclidean spaces with the fixed coordinate systems $x = (x_1, \dots, x_N) \in \mathbb{R}^N$ and $a = (a_1, \dots, a_k) \in \mathbb{R}^n$ with $N \geq n$ and a nontrivial vector-polynomial $P \colon \mathbb{R}^N \to \mathbb{R}^n$. Recall that P is nontrivial if the image $P(\mathbb{R}^N)$ has nonempty interior. For brevity in what follows we use *polynomial* for *vector-polynomial.*

DEFINITION 19. Let $m \in \mathbb{Z}_+$ and $\delta > 0$. We call the (m, δ)-cone $K^n_{m,\delta}$ the following set of points:

$$K^n_{m,\delta} = \{a = (a_1, \dots, a_N) \in \mathbb{R}^N : 0 < |a_1| < \delta, 0 < |a_{j+1}| < |a_1 \cdots a_j|^m \text{ for } j = 1, \dots, N - 1\}. \tag{48}$$

Let $m' \in \mathbb{Z}_+^N$. If $m' \geq m$ and $\delta' \leq \delta$, then we say that the (m', δ')-cone $K^n_{m',\delta'}$ is a refinement of the (m, δ)-cone $K^n_{m,\delta}$.

Define the following sets:

$$V_{m,\delta,P} = \text{closure}\{P^{-1}(K^n_{m,\delta})\}, \quad V_{0,m,P} = \bigcap_{\delta > 0} V_{m,\delta,P} \tag{49}$$

Then one has

THEOREM 9 ([**K6**]). *For any nontrivial polynomial* $P\colon \mathbb{R}^N \to \mathbb{R}^n$ *there exist an integer* $m \in \mathbb{Z}_+$ *and a positive* δ *such that the following conditions hold*

(a) *the set* $V_0 = V_{0,m,P}$ *(see (49)) is a semialgebraic set of codimension at least* n.

(b) *the set* $V_{m,\delta,P}$ *consists of regular points of* P*, i.e., if* $b \in V_{m,\delta,P}$*, then the level set* $P^{-1}\big(P(b)\big)$ *is a manifold of codimension* n.

(c) *there exists a stratification of* V_0 *by semialgebraic strata* $(V_0, \mathcal{V}_0)$ *satisfying the property*: $V_{m,\delta,P}$ *is* a_P*-regular over any strata* $V_i \in \mathcal{V}_0$ *with respect to* P.

REMARK 7. In order to have the compatibility condition for the limiting set $V_{m,\delta,P}$ with the regular level sets $P^{-1}(a)$ in the definition of the (m,δ)-cone $K^n_{m,\delta}$ (48) it is necessary that the range of values ("smallness") of a_{j+1} depends on all a_i's with $i = 1, \dots, j$. Indeed, consider the following

EXAMPLE 5. Let $x = (x_1, x_2, x_3)$ denote a point in $\mathbb{R}^3$ and $P = (P_1, P_2, P_3)\colon \mathbb{R}^3 \to \mathbb{R}^3$ be a polynomial map, given by

$$P_1(x) = x_1, \quad P_2(x) = x_1x_2, \quad P_3(x) = x_1x_2x_3. \tag{50}$$

If the definition of the (m, δ)-cone is

$$K^3_{m,\delta} = \{a = (a_1, a_2, a_3) \in \mathbb{R}^3 : 0 < |a_1| < \delta,\ 0 < |a_{2(\text{resp. }3)}| < |a_1|^m\}, \tag{51}$$

then the limiting set $V_{m,\delta,P}$, defined by (49), is 1-dimensional for any positive m. However, all level sets $P^{-1}(a)$ with $a_1a_2 \neq 0$ are 0-dimensional. In this case the compatibility condition in Definition 18 fails.

3.4. A proof of Hironaka's theorem on existence of a_P-stratifications for polynomial functions. In this section we present a geometric proof of Hironaka's theorem based on a proof of Whitney's Theorem 7 on existence of a-stratifications due to the author [**K7**]. First, we briefly outline the latter proof and then prove Hironaka's theorem following the same path.

3.4.1. *An outline of a proof of Whitney's Theorem 7 on existence of* a*-stratifications.* The outline given below works to prove b-stratifications too after a slight modification [**K7**].

A semivariety V has well-defined dimension, say $d \leq N$. Denote by V_{reg} the set of points where V is locally a real algebraic submanifold of $\mathbb{R}^N$ of dimension d. V_{reg} is a semivariety, moreover, $V_{\text{sing}} = V \setminus V_{\text{reg}}$ is a semivariety of positive codimension in V, i.e., $\dim V_{\text{sing}} < \dim V$. (For the algebraic case see, e.g., [**M3**].)

Step 1. There is a filtration of V by semivarieties

$$V^0 \subset V^1 \subset \cdots \subset V^d = V, \tag{52}$$

where for each $k = 1, \dots, d$ the set $V^k \setminus V^{k-1}$ is a manifold of dimension k. This is not difficult to see, e.g., [**M3**]. Indeed, consider $V_{\text{sing}} \subset V$, then $V \setminus V_{\text{sing}}$ is a manifold of dimension d and $\dim V_{\text{sing}} < d$. Inductive application of these arguments completes the proof.

A *refinement* of a decomposition $V = \bigsqcup_{i \in I} V_i$ is a decomposition $V = \bigsqcup_{i' \in I'} V_{i'}$ such that any stratum V_j of the first decomposition is a union of some strata of the second one, i.e., there is a set $I'(j) \subset I'$ such that $V_j = \bigsqcup_{i' \in I'(j)} V_{i'}$.

Step 2. Let $V \subset \mathbb{R}^N$ be a manifold and $W \subset V$ be a semivariety. Denote by $\mathrm{Int}_V(W)$ the set of interior points of W in V w.r.t. the topology induced from $\mathbb{R}^N$. Let V_i and V_j be a pair of distinct strata. For each point $x \in V_i \cap \overline{V_j}$ denote by $V_j^{\mathrm{con},x}$ the local connected component of V_j at x, i.e., the connected component of the intersection of V_j with a small ball centered at x and call it *essential* if the closure of $V_j^{\mathrm{con},x}$ has x in the interior, $x \in \mathrm{Int}_{V_i}(V_i \cap \overline{V_j^{\mathrm{con},x}})$. Denote by $V_j^{\mathrm{ess},x}$ the union of all local essential components of V_j. A semialgebraic set V_j can have only finitely many local connected components (see, e.g., [**M3**]).

THEOREM 10. *For any two disjoint strata V_j and V_i the set of points*

$$\mathrm{Sing}_{\mathrm{a}}(V_j, V_i) = \{x \in V_i \cap \overline{V_j} : (V_j^{\mathrm{ess},x}, V_i, x) \text{ is not a-regular}\},$$

is a semivariety in V_i and $\dim \mathrm{Sing}_{\mathrm{a}}(V_j, V_i) < \dim V_i$.

Let us show that this theorem is sufficient to prove the a-regular case of Theorem 7. Consider a decomposition $V = \bigsqcup_{i \in I} V_i$ and split the strata into two groups: the first group consists of strata of dimension at least k and the second group consists of the rest. Suppose that each stratum from the first group is a-regular over each stratum from the second group. Then by definition of a-regularity any refinement of a stratum from the second group preserves this a-regularity.

Now apply this refinement inductively. Consider strata in $V^d \setminus V^{d-1}$ of dimension d. Using Theorem 10 and the result of Łojasiewicz [**Ł1**] that the frontier of a semivariety has dimension less than the semivariety itself, refine V^{d-1} so that each d-dimensional stratum is a-regular over each stratum in V^{d-1}. The above remark shows that any further refinement of the strata in V^{d-1} preserves the a-regularity of strata from $V^d \setminus V^{d-1}$ over it. This reduces the problem of existence of stratification for d-dimensional semivarieties to the same problem for $(d-1)$-dimensional semivarieties. Induction on dimension completes the proof of Theorem 7.

Our proof is based on the observation that if $V_i \subset \overline{V_j}$ are a pair of strata, a-regularity of V_j over V_i at x in V_i is closely related to whether the limit of tangent planes $T_y V_j$ is unique or not as y from V_j tends to x. The rest of the paper is devoted to the proof of Theorem 10, which consists of two steps. In Lemma 5 we relate a-regularity with (non)uniqueness of limits of tangent planes $T_y V_j$; then, based on it and Rolle's lemma in Lemma 6, one can prove Theorem 10.

Let V_i and V_j be a pair of distinct strata in $\mathbb{R}^N$. Define

$$\begin{aligned}\mathrm{Un}(V_j, V_i) = \{x \in V_i \cap \overline{V_j} : &\text{ for any } V_j^{\mathrm{con},x}, \text{ there exists } \tau_x \subset T_x\mathbb{R}^N \\ &\text{such that for any } \{y_n\} \subset V_j^{\mathrm{con},x} \text{ tending to } x,\ T_{y_n} V_j \to \tau_x\},\end{aligned} \tag{53}$$

The proof consists of two lemmas.

LEMMA 5. *With the above notations we have*

$$\mathrm{Sing}_{\mathrm{a}}(V_j, V_i) \subset V_i \setminus \mathrm{Un}(V_j, V_i), \tag{54}$$

where $\mathrm{Sing}_a(V_j, V_i)$ *is the set of* a-*singular points in V_i defined in Theorem* 10.

LEMMA 6. *With the above notations there is a set of strata* $\{V_j^p\}_{p\in\mathbb{Z}}$ *(resp.* $\{V_i^p\}_{p\in\mathbb{Z}}$*) in* V_j *(resp. in* V_i*), each of positive codimension in* V_j *(resp. in* V_i*), such that*

$$(55)\qquad \begin{aligned} \mathrm{Sing}_{\mathrm{a}}(V_j, V_i) &\subset \bigcup_{p\in\mathbb{Z}} \mathrm{Sing}_{\mathrm{a}}(V_j^p, V_i) \bigcup_{p\in\mathbb{Z}} V_i^p. \\ V_i \setminus \mathrm{Un}(V_j, V_i) &\subset \bigcup_{p\in\mathbb{Z}} V_i \setminus \mathrm{Un}(V_j^p, V_i) \end{aligned}$$

REMARKS. (1) Inductive application of this lemma to the right-hand side of the first line of (55) reduces dimensions of V_j^p's up to $\dim V_i$.

(2) The dimension of the frontier of a semivariety ($\mathrm{Sing}_{\mathrm{a}}(V_j^p, V_i) \subset V_i \cap \overline{V_j^p}$) has dimension strictly smaller than that of the semivariety (V_j^p) itself.

(3) By Lemma 5 the set $\mathrm{Sing}_{\mathrm{a}}(V_j, V_i)$ is a semivariety. Since a countable union of semivarieties of positive codimension in V_i contains $\mathrm{Sing}_{\mathrm{a}}(V_j, V_i)$ we have $\mathrm{Sing}_{\mathrm{a}}(V_j, V_i)$ has positive codimension in V_i, which proves Theorem 10.

Since this proves Theorem 10, as a consequence this proves Theorem 7 too. We are not going to prove these lemmas; however, we would like to exhibit the geometric idea behind the proof of Lemma 6. The section below is devoted to the idea of construction of proper subvarieties in the bigger stratum V_j approaching the non-a-regular points $\mathrm{Sing}_{\mathrm{a}}(V_j, V_i) \subset V_i$.

3.4.2. *Separation of Planes and dimension reduction in Lemma* 6. Let τ_0 and τ_1 be two distinct oriented k-dimensional planes in $\mathbb{R}^N$. An oriented $(m-k)$-dimensional plane l in $\mathbb{R}^N$ *separates* τ_0 and τ_1 if l is transverse to τ_0 and τ_1 and the orientations induced by $\tau_0 + l$ and $\tau_1 + l$ in $\mathbb{R}^N$ are different. Notice that there always exists an open set of orientable $(m-k)$-planes separating any two distinct orientable k-planes.

ROLLE'S LEMMA. *If a continuous family of orientable k-planes* $\{\tau_t\}_{t\in[0,1]}$ *connects* τ_0 *and* τ_1 *and an orientable* $(m-k)$*-plane* l *separates* τ_0 *and* τ_1*, then for some* $t^* \in (0,1)$ *transversality of* τ_{t^*} *and* l *fails.*

In what follows we use the transversality theorem [**GG**] which says: *if* $V \subset \mathbb{R}^N$ *is a manifold, then almost every plane of dimension* k *is transverse to* V.

AN OUTLINE OF THE PROOF OF LEMMA 6. Let $x \in \mathrm{Sing}_{\mathrm{a}}(V_j, V_i)$, then by Lemma 5 there are sequences $\{y'_n\}, \{y_n\} \subset V_j^{\mathrm{con},x}$ with different limiting tangent planes $\tau = \lim T_{y_n} V_j$ and $\tau' = \lim T_{y'_n} V_j$. Choose an orientation of $T_{y_0} V_j$. By connecting y_0 locally with all other points $\{y'_n\}$ one can induce an orientation on all other $T_{y'_n} V_j$ so that the orientations of τ_0 and τ_1 coincide with the orientations of the limits.

Denote $\dim V_j$ by k. There is an oriented $(N-k)$-plane l_j separating τ_0 and τ_1 and transverse to V_j (by the transversality theorem). Consider the orthogonal projection π_{l_j} along l_j onto its orthogonal complement $l_j^{\perp}$. Denote by $p_{l_j,j}$ its restriction to V_j, $p_{l_j,j} = \pi_{l_j}|_{V_j} : V_j \to l_j^{\perp}$. Denote by $\mathrm{Crit}(l_j, V_j)$ the set of critical points of $p_{l_j,j}$ in V_j where the rank of $p_{l_j,j}$ is not maximal. Then $\mathrm{Crit}(l_j, V_j)$ is a semivariety in V_j and $\dim \mathrm{Crit}(l_j, V_j) < \dim V_j$. Connect two points y_n^0 and y_n^1 by a curve in V_j; then $T_{y_n^0} V_j$ deforms continuously to $T_{y_n^0} V_j$. Then by Rolle's lemma there is a critical point of $p_{l_j,j}$ in $V_j^{\mathrm{con},x}$ arbitrarily close to x. Thus $x \in \overline{\mathrm{Crit}(l_j, V_j)}$.

By the transversality theorem there is a countable dense set of orientable $(N-k)$-planes $\{l_j^p\}_{p\in\mathbb{Z}_+}$ transverse to V_j and separating any two distinct orientable k-planes τ_0 and τ_1. Therefore, we have that

$$\text{(56)} \qquad \mathrm{Sing}_{\mathrm{a}}(V_j, V_i) \subset V_i \setminus \mathrm{Un}(V_j, V_i) \subset \bigcup_{p\in\mathbb{Z}_+} \{\overline{\mathrm{Crit}(l_j^p, V_j)} \setminus \mathrm{Crit}(l_j^p, V_j)\}.$$

These sets $\{\mathrm{Crit}(l_j^p, V_j) = V_j^p\}_{p\in\mathbb{Z}_+}$ are the proper subsets in V_j we are looking for. Using some additional simple argument given in [**K7**] one can complete the proof of Lemma 6. □

3.4.3. *A proof of* a_P*-stratifications for polynomial functions.* The proof below also consists of two steps.

Step 1. *Construct a rank stratification of P.* Consider an a-regular stratification $(V, \mathcal{V}^0)$ of V by semialgebraic strata. Such a stratification always exists by Whitney's Theorem 7 proved above. Now we refine the stratification $\mathcal{V}^0$ to a stratification $\mathcal{V}^1$ so that $\mathcal{V}^1$ is a rank stratification of P or the restriction of P to any stratum $V_i \in \mathcal{V}^1$ is a map of constant rank. Notice that it is sufficient to refine each stratum $V_i \subset \mathcal{V}^0$ so that P restricted to each stratum $V_i^j \subset V_i$ has constant rank.

There are two cases: if $P(V_i)$ is a point, then the rank of $P|_{V_i}$ is identically zero and V_i stays unchanged and if $P(V_i)$ contains an open set, then denote by $\Sigma_{i,P} \subset V_i$ the set of critical points of $P|_{V_i}$. By Sard's lemma for algebraic sets [**M**] the set $\Sigma_{i,P}$ is a semialgebraic set of positive codimension in V_i. Now refine each $\Sigma_{i,P}$ to be an a-regular stratification of $\Sigma_{i,P}$. This is possible by Whitney's Theorem 7. Denote such a stratification by $(V, \mathcal{V}^1)$. By our construction $\mathcal{V}^1$ is an a-regular rank stratification of $P|_V$, i.e., P has constant rank on each stratum and the strata "fit" a-regularly.

Step 2. It is sufficient to prove the following

THEOREM 11. *Let $V_i, V_j \subset \mathcal{V}^1$ be two strata in $\mathbb{R}^N$ and $P\colon \mathbb{R}^N \to \mathbb{R}$ be a constant on each strata V_i and V_j. Then the set of singular points*

$$\mathrm{Sing}_{\mathrm{a},P}(V_j, V_i) = \{x \in V_i \cap \overline{V_j} : V_j \textit{ is not } \mathrm{a}_P\textit{-regular over } V_i \textit{ at } x \textit{ w.r.t. } P\}$$

is semialgebraic and has positive codimension in V_i.

Inductive refinement arguments from Section 3.4.1 along with Theorem 11 complete the proof of Hironaka's Theorem 8. The rest of the section is devoted to a proof of Theorem 11.

PROOF OF THEOREM 11. Similarly to the existence proof of Whitney's Theorem 7 above we define the set with a unique limit of tangent planes to level sets of P

$$\text{(57)} \quad \mathrm{Un}_P(V_j, V_i) = \{x \in V_i : \text{for any } V_j^{\mathrm{con},x} \text{ there is } \tau_{x,P} \text{ such that } \lim_{y_n\to x} \ker dP|_{V_j^{\mathrm{con},x}}(y_n) = \tau_{x,P} \text{ and is unique}\}.$$

LEMMA 7. *With the above notations* $\mathrm{Un}_P(V_j, V_i)$ *and* $\mathrm{Sing}_{\mathrm{a},P}(V_j, V_i)$ *are semivarieties and*

$$\text{(58)} \qquad \mathrm{Sing}_{\mathrm{a},P}(V_j, V_i) \subset V_i \setminus \mathrm{Un}_P(V_j, V_i).$$

LEMMA 8. *With the above notations there is a set of strata* $\{V_j^p\}_{p\in\mathbb{Z}}$ *(resp.* $\{V_i^p\}_{p\in\mathbb{Z}}$*) in* V_j *(resp. in* V_i*) each of positive codimension in* V_j *(resp. in* V_i*) such that*

$$\mathrm{Sing}_{\mathrm{a},P}(V_j, V_i) \subset \bigcup_{p\in\mathbb{Z}} \mathrm{Sing}_{\mathrm{a},P}(V_j^p, V_i) \bigcup_{p\in\mathbb{Z}} V_i^p. \tag{59}$$

REMARK 8. Similarly to the remarks after Lemma 6 this lemma allows one to reduce the dimension of V_j's and prove that $\mathrm{Sing}_{\mathrm{a},P}(V_j, V_i)$ has positive codimension in V_i. This would prove Theorem 11 and as a consequence it would prove Theorem 8. So what is left to prove is Lemmas 7 and 8.

PROOF OF LEMMA 7. The proof goes by contradiction. Let $x \in \mathrm{Un}_P(V_j, V_i) \cap \mathrm{Sing}_{\mathrm{a},P}(V_j, V_i)$. Then for any local connected component $V^{\mathrm{con},x}$ and any sequence $\{y_n\} \subset V^{\mathrm{con},x}$ there is a limiting plane $\lim \ker dP|_{V_j}(x) = \tau_{x,P}$. Moreover, we have $\ker dP|_{V_i}(x) \not\subset \tau_{x,P}$. Thus, there is a unit vector $v \in \ker dP|_{V_i}$ and $v \notin \tau_{x,P}$. The contradiction we are going to get is to find a sequence of points $\{y_n\} \subset V^{\mathrm{con},x}$ such that $\lim \ker dP|_{V_j}(y_n) = \tau' \supset v$. The rest of the proof is devoted to the construction of such a sequence.

By the theorem on implicit functions one can straighten V_i along with the nonempty level sets $P^{-1}(a) \cap V_i$. Then the ray $l_v(x) = \{y \in \mathbb{R}^N : (y-x)/|y-x| = v\} \subset P^{-1}\big(P(x)\big) \subset V_i$ belongs to the level set $P^{-1}\big(P(x)\big)$. By an extension of Wall [**W1**] of Milnor's curve selection lemma there is a 2-dimensional "wing" $V_{j,v} \subset V_j$ such that $l_v(x) \subset \overline{V_{j,v}}$.

By Lemma 6 the set of points with nonunique limit $\mathrm{Un}\big(V_{j,v}, l_v(x)\big)$ is 0-dimensional. Therefore, by Lemma 5 there is a neighborhood U_x of x such that any $y \in U_x \cap l_v(x)$ may be distinct from x, $V_{j,v}$ is a-regular over $l_v(x)$ at y and the limit τ_y of tangent planes $T_{y_n}V_{j,v}$ as $y \to x$ is unique. The last two properties imply that τ_y depends continuously on y as long as the limit τ_y is unique. Therefore, there is a neighborhood $U_y \subset U_x \setminus x$ of y such that $V_{j,v} \cap U_y$ is a C^1-manifold with boundary.

Consider a C^1-smooth one-sided chart in $U_y \cap V_{j,v}$ and let π_L be the map from $U_y \cap V_{j,v}$ into the 2-dimensional plane $\mathbb{R}^2$. The image $\pi_L(V_i)$ is a line in $\mathbb{R}^2$ and $\pi_L(V_{j,v})$ is a one-sided neighborhood of this line. Using Rolle's type of argument it is easy to show this for a sequence $\{y_n\} \subset \pi_L(V_{j,v})$ from a semineighborhood of $\pi_L(x') \in \mathbb{R}^2$ such that $y_n \to \pi_L(y)$ and $T_{y_n'}\{\pi_L \circ P^{-1}(P(\pi_L^{-1}(y_n')))\} \to \pi_L(v)$. This implies that $\lim \ker dP|_{V_{j,v}}(y_n')$ tends to v, however, $V_{j,v} \subset V_j^{\mathrm{con},x}$. So $v = \lim_{n\to\infty} \ker dP|_{V_{j,v}}(\pi_L^{-1}(y_n)) \subset \lim_{n\to\infty} \ker dP|_{V_j^{\mathrm{con},x}}(\pi_L^{-1}(y_n)) = \tau$. This is a contradiction with $v \notin \tau$. □

PROOF OF LEMMA 8. The proof is almost the same as the proof of Lemma 6 outlined above [**K7**]. □

This completes the proof of Theorem 11 of Hironaka. □

Acknowledgments. I would like to thank my thesis advisor John Mather and David Nadler for stimulating discussions and numerous remarks on the mathematics of this lecture.

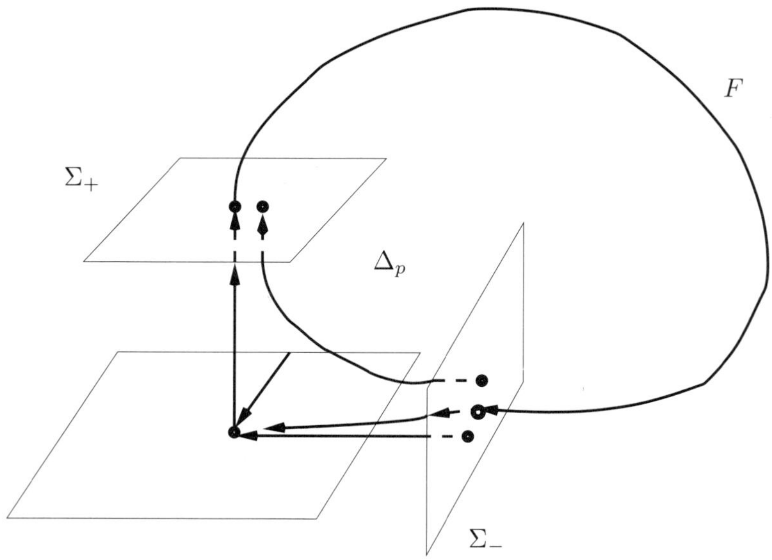

FIGURE 10. A saddle loop polycycle

4. Bifurcation of spatial polycycles and blow-up along the diagonal of the space of multijets

In this lecture we discuss an essential ingredient of the proof of Theorem 1 [**K8**] about an estimate on cyclicity of spatial quasielementary polycycles. First, in Section 4.1 we motivate the appearance of multichain maps (38) to get an estimate on cyclicity of spatial polycycles. Similarly to the planar case the question of estimating cyclicity of a quasielementary polycycle reduces to estimating the geometric multiplicity of a multichain map of the form (38). To get an estimate on geometric multiplicity of a multichain map (38) one needs to prove a Bézout type theorem for multichain maps. However, a straightforward way to prove a Bézout type theorem for multichain maps faces a typical problem in singularity theory, namely, the problem that transversality fails on the diagonal in the space of multijets (see, e.g., [**GG, M2**]). We shall overcome this problem using a construction of Grigoriev–Yakovenko [**GY**] of blow-up along the diagonal in the space of multijets and a special multijet transversality theorem [**GY**]. This construction and its relation to Newton Interpolation Polynomials are described in Section 4.2. In Section 4.3 we describe the problem of rate of growth of the number of periodic points from Smooth Dynamical Systems (see, e.g., [**AM, S4**]) and outline the main result of the author along with Brian Hunt [**KH, K5**] in this direction. Finally, in Section 4.4 we outline how Newton Interpolation Polynomials can be applied to perturb trajectories and control the number of periodic points of diffeomorphisms.

4.1. Multichain maps and spatial polycycles. Consider the simplest example of a polycycle γ in $\mathbb{R}^3$ consisting of a saddle equilibrium p and a connecting separatrix γ_p (see Figure 10).

Let Σ^- and Σ^+ be "entrance" and "exit" transverse sections to γ_p chosen so that in C^r-normal coordinates the Poincaré return map Δ_p along the polycycle γ has a "nice" form. We decompose the Poincaré return map Δ along the polycycle

γ into the composition of a local Poincaré map Δ_p in a neighborhood of p and a semilocal map f along γ_p. Consider 2-cycles bifurcating from γ. Denote by x_1, x_2 and y_1, y_2 the first and the second intersection of each one of the 2-cycles with Σ^+ and Σ^- respectively. Then the equation determining the number of 2-cycles has the form

$$\begin{cases} y_1 = f(x_1, \varepsilon) \\ x_2 = \Delta(y_1, \varepsilon) \\ y_2 = f(x_2, \varepsilon) \\ x_1 = \Delta(y_2, \varepsilon). \end{cases} \tag{60}$$

It is important that the first and the third equation have the same functional parts. Notice also that each of the equations from (60) is an equality in $\mathbb{R}^2$ so it itself consists of two 1-dimensional equalities. So the total number of equations in (60) is 8. Following the strategy of the planar case from Section 2.2 we apply the Khovanskiĭ method to the system (60). Compare this system with the system (21) or (23). It is not difficult to see that the result of application of the Khovanskiĭ method gives a map of the form

$$P \circ (j^7 f, j^7 f)\colon \mathbb{R}^8 \to \mathbb{R}^8. \tag{61}$$

To simplify our considerations we redenote the seventh jet $j^7 f$ of f by a map F and consider the multichain map

$$P \circ (F, F)\colon \mathbb{R}^8 \to \mathbb{R}^8, \tag{62}$$

where F is a *generic map* in a sense that it satisfies *any pre-given transversality condition.* It is clear that even if F is generic we cannot assume that the 2-tuple (F, F) is a generic map, simply because the first and the second components are the same. Let's explain why genericity fails for a 2-tuple mapping by an example.

4.1.1. *Genericity (resp. transversality) fails for 2-tuple mappings!*

CLASSICAL TRANSVERSALITY THEOREM (e.g., [**AGV, GG**]). *Let N and m be positive integers and M be a smooth compact manifold in $\mathbb{R}^m$. Then for an open dense set of smooth mappings $F\colon \mathbb{R}^N \to \mathbb{R}^m$ we have that F is transverse to M. In particular, this means that $F^{-1}(M)$ is a smooth manifold.*

REMARKS. (1) It is an exercise from calculus to construct a set on the unit interval $[0, 1]$ which is open dense and has arbitrary small positive measure. To justify that the transversality property is indeed generic, there is the so-called prevalent extension of the classical transversality theorem which says that for an a.e. mapping $F\colon B^n \to \mathbb{R}^N$ we have that F is transverse to M. More exactly, for a generic finite-parameter family of mappings $\{F_\varepsilon\colon B^n \to \mathbb{R}^N\}_{\varepsilon \in B^k}$ for a.e. parameter value, F_ε is transverse to M. See [**HSY, K1**] for more details.

(2) The fact that transversality of F to M implies that $F^{-1}(M)$ is a smooth manifold follows from the theorem on implicit functions (see [**AGV, GG**]).

Since for a genericity condition on F what we need is that F has to satisfy a transversality condition, to show the impossibility of application of the classical transversality theorem we give a trivial example when transversality fails for an open set of 2-tuples (F, F).

EXAMPLE 6. In the classical transversality theorem put $n = m = 1$. Consider the function $f\colon x \to x^2$ for $x \in I = [-1, 1]$ and the corresponding 2-tuple $f \times f\colon I \times I \to \mathbb{R} \times \mathbb{R}$, given by $f \times f\colon (x_1, x_2) \to (x_1^2, x_2^2) = (y_1, y_2)$. Let $M = \{y_1 = y_2\} \subset \mathbb{R} \times \mathbb{R}$ be the diagonal. Then for each $\tilde{f}$ which is C^1-close to f the preimage $(\tilde{f}, \tilde{f})^{-1}(M)$ is a topological cross (not a manifold). This, in particular, implies that $(\tilde{f}, \tilde{f})$ is not transverse to M, otherwise the preimage of a manifold would be a manifold.

To see this, notice that a function $\tilde{f}$ close to f has to have a local minimum $\tilde{x}$ close to 0 and $\tilde{x}$ is a nondegenerate local minimum, i.e., $\tilde{f}\colon x \to \tilde{\varepsilon} + \tilde{a}(x - \tilde{x})^2 + o\big((x - \tilde{x})^2\big)$ with $\tilde{a} \neq 0$. Then $\tilde{f}(x_1) - \tilde{f}(x_2) = 0$ has two intersecting curves of solutions $x_1 = x_2$ and $x_1 - \tilde{x} \approx -(x_2 - \tilde{x})$ which form a cross. This completes the proof of the claim in the example.

To explain what happens in this example we derive a general framework due to Grigoriev–Yakovenko [**GY**].

4.1.2. *Blow-up along the diagonal for 2-tuples in the 1-dimensional case.* For a smooth function $\tilde{f}\colon \mathbb{R} \to \mathbb{R}$ consider the maps

$$(63)\qquad \begin{aligned} (x_1, x_2) &\xrightarrow{\mathcal{D}_2\tilde{f}} \left(x_1, x_2, \tilde{f}(x_1), \frac{\tilde{f}(x_2) - \tilde{f}(x_1)}{x_2 - x_1}\right) = (x_1, x_1, u_1, u_2) \subset \mathbb{R}^4, \\ (x_1, x_2, u_1, u_2) &\xrightarrow{\pi_2} \big(x_1, x_2, u_1, u_1 + u_2(x_2 - x_1)\big). \end{aligned}$$

Direct calculation shows that $\pi_2 \circ \mathcal{D}_2\tilde{f} \equiv (\tilde{f}, \tilde{f})$. Therefore,

$$(64)\qquad (\tilde{f}, \tilde{f})^{-1}(M) = (\mathcal{D}_2\tilde{f})^{-1} \circ \pi_2^{-1}(M).$$

This is incorporated into Figure 3 of with $n = N = 1$. By definition $\mathcal{D}_2(f)\colon I \times I \to \mathcal{D}\mathcal{D}_2(I, \mathbb{R})$ is a smooth map, $\pi_2\colon \mathcal{D}\mathcal{D}_2(I, \mathbb{R}) \to \mathbb{R}$ is an explicitly computable polynomial map, and $\pi_2 \circ \mathcal{D}_2(f) = (f, f)\colon I \times I \to \mathbb{R}^2$. Notice that outside of the diagonal $\{x_1 = x_2\}$ the map π_2 is one-to-one. However, the preimage of the set $\pi_2^{-1}\{x_1 = x_2, f(x_1) = f(x_2)\}$ is of dimension 3 while the set $\{x_1 = x_2, f(x_1) = f(x_2)\}$ itself is of dimension 2. This, in particular, means that π_2 is a blow-up along the diagonal.

Consider $\pi_2^{-1}(M) = \{u_2(x_2 - x_1) = 0\} \subset \mathbb{R}^4$. This is the union of two intersecting hyperplanes. If the map $\mathcal{D}_2\tilde{f}$ is transverse to $\{u_2(x_2 - x_1) = 0\}$, then the preimage $(\mathcal{D}_2\tilde{f})^{-1}(\{u_2(x_2 - x_1) = 0\})$ has to be the union of two intersecting curves. It turns out that the assumption that $\mathcal{D}_2\tilde{f}$ is *generic for a generic* $\tilde{f}$ is satisfied or for a generic $\widetilde{F}$ the map $\mathcal{D}_2\tilde{f}$ is transverse to both hyperplanes of $\{u_2(x_2 - x_1) = 0\}$. Let's justify that.

A "PROOF" OF THE CLASSICAL TRANSVERSALITY THEOREM OF THOM (see, e.g., [**AGV, GG**]). Transversality is an open property, i.e., if F is transverse to M, then for all $\widetilde{F}$ sufficiently close to F we have $\widetilde{F}$ is transverse to M too. So it is sufficient to show that by an arbitrary small perturbation of any mapping $F\colon \mathbb{R}^N \to \mathbb{R}^m$ one can reach transversality to a compact manifold M. Let's prove it now.

Consider a smooth mapping $F\colon B^n \to \mathbb{R}^N$. Include this mapping into the m-parameter family $\mathbf{F}\colon \mathbb{R}^n \times \mathbb{R}^N \to \mathbb{R}^n \times \mathbb{R}^N$, given by $\mathbf{F}(x, \varepsilon) = (x, F(x) + \varepsilon)$. The determinant of the linearization of the mapping (the Jacobian) $J_{\mathbf{F}}(x, \varepsilon)$ is constant and identically equals 1. Therefore, $\mathbf{F}$ is a diffeomorphism and $M_F = \mathbf{F}^{-1}(\mathbb{R}^n \times M)$ is a manifold in the preimage $\mathbb{R}^n \times \mathbb{R}^N$.

FACT. *If ε is a regular point of the projection $\pi_{M,F} = \pi|_{M_F} : \mathbb{R}^n \times \mathbb{R}^N \to \mathbb{R}^N$ along the x-coordinate, restricted to M_F, then $F_\varepsilon = F(x) + \varepsilon$ is transverse to M.*

This follows from the implicit function theorem.

SARD'S LEMMA (e.g., [**M4**]). *A.e. $\varepsilon \in \mathbb{R}^N$ is a regular value for the projection map $\pi_{M,F}$.*

Thus, one can choose a regular value ε arbitrarily close to 0. For such an ε the mapping F_ε is transverse to M. This completes the proof of the classical transversality theorem. □

Now we are ready to state the main result of this section.

4.1.3. *Multijet transversality theorem due to Grigoriev – Yakovenko.*

THEOREM 12 ([**GY**]). *Let $M \subset \mathbb{R}^N \times \mathbb{R}^N$ be a smooth variety and $B^n \subset \mathbb{R}^n$ be a unit ball. Then for an open dense set of smooth mappings $F\colon B^n \to \mathbb{R}^N$ the set $(F \times F)^{-1}(M)$ is stratified.*

Moreover, let $k \in \mathbb{Z}_+$ and $M \subset \mathbb{R}^N \times \cdots \times \mathbb{R}^N$ (k times) be a smooth algebraic variety. Then for an open dense set of smooth mappings $F : B^n \to \mathbb{R}^N$ the set $(F \times \cdots \times F)^{-1}(M)$, with k repetitions, is stratified.

Moreover, let $n, k \in \mathbb{Z}_+$ and $M \subset J^m(B^n, \mathbb{R}^N) \times \cdots \times J^m(B^n, \mathbb{R}^N)$ (k times) be a smooth algebraic variety. Then for an open dense set of smooth mappings $F\colon B^n \to \mathbb{R}^N$ the set $(F \times \cdots \times F)^{-1}(M)$, with k repetitions, is stratified.

PROOF OF THE THEOREM FOR THE MODEL EXAMPLE $n = m = 1$ AND $k = 2$. Consider the map $\mathcal{D}_2(F)\colon (x_1, x_2, \varepsilon_1, \varepsilon_2) \mapsto (x_1, x_2, u_1, u_2)$, defined by the formula (63). Direct calculations show that the determinant of the linearization (the Jacobian) $J_{\mathcal{D}_2(F)}(x_1, x_2, \varepsilon_1, \varepsilon_2) \equiv 1$ and is formed by an upper triangular matrix with units on the diagonal. Since, $\mathcal{D}_2(F)$ is a diffeomorphism one can apply arguments of the proof of the classical transversality theorem given above. □

A proof of the theorem in the general case follows along the same lines. The main difficulty is to construct Figure 3 in the general case. This is the subject of the next subsection.

COROLLARY 3. *For an open dense set of smooth functions $F\colon I \to \mathbb{R}$ the preimage $(F, F)^{-1}(M)$ is a 1-dimensional stratified manifold, i.e., a locally finite union of points and curves.*

4.2. Newton interpolation polynomials and blow-up along the diagonal in the space of multijets. This section is devoted to the description of the Grigoriev – Yakovenko construction of blow-up along the diagonal in the space of multijets in the general case. Let $F\colon B^n \to \mathbb{R}^N$ be a smooth map of a unit ball $B^n \subset \mathbb{R}^n$, $j^m F\colon B^n \to J^m(B^n, \mathbb{R}^N)$ be an mth-order jet of F, and $\mathcal{J}^{m,k}(B^n, \mathbb{R}^N) = J^m(B^n, \mathbb{R}^N) \times \cdots \times J^m(B^n, \mathbb{R}^N)$ (k repetitions) be the space of k-tuple m-jets. Denote the k-tuple of the m-jet of a map $F\colon \mathbb{R}^N \to \mathbb{R}^m$ by $\mathcal{J}^{m,k} F(x_1, \dots, x_k) = (j^m F(x_1), \dots, j^m F(x_k))$. The goal of this section is to define entries of an extension of Figure 3: the so-called space of divided differences $\mathcal{DD}_k^m(B^n, \mathbb{R}^N)$, the Newton map $\pi_k^m\colon \mathcal{DD}_k^m(B^n, \mathbb{R}^N) \to \mathcal{J}^{m,k}(B^n, \mathbb{R}^N)$, $\mathcal{D}_k^m(F)\colon \underbrace{B^n \times \cdots \times B^n}_{k \text{ repetitions}} \to \mathcal{DD}_k^m(B^n, \mathbb{R}^N)$. We use the exposition from [**GY**].

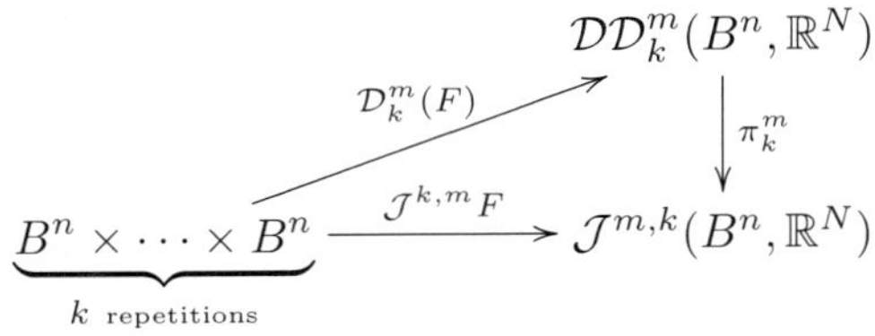

FIGURE 11. Polynomial blow-up of the multijet space

4.2.1. *Divided differences.* In order to extend the above construction we need to define so-called *divided differences*. Let $g\colon \mathbb{R} \to \mathbb{R}$ be a sufficiently smooth function of one real variable.

DEFINITION 20. The *first order divided difference* of g is defined as

$$\Delta g(x_1, x_2) = \frac{g(x_2) - g(x_1)}{x_2 - x_1} \tag{65}$$

for $x_2 \neq x_1$ and extended by its limit value as $g'(x)$ for $x = x_2 = x_1$. Clearly, if g is a C^r-smooth function, then Δg is at least a C^{r-1}-smooth function of its arguments.

Iterating this construction we define *divided differences of the sth order for any* $s \in \mathbb{Z}_+$ as

$$\Delta^s g(x_1, \ldots, x_{s+1}) = \frac{\Delta^{s-1} g(x_1, \ldots, x_{s-1}, x_{s+1}) - \Delta^{s-1} g(x_1, \ldots, x_{s-1}, x_s)}{x_{s+1} - x_s}$$

for $x_{s+1} \neq x_s$ and extended by its limit value as $\partial \Delta^{s-1} g(x)/\partial x_s$ for $x = x_{s+1} = x_s$. Clearly, if g is a C^r-smooth function, then Δg is at least a C^{r-s}-smooth function of its arguments.

Notice that Δ^s is linear as a function of g, and one can show that it is a symmetric function of $x_1, \ldots, x_{s+1}$; in fact, by induction it follows that

$$\Delta^s g(x_1, \ldots, x_s) = \sum_{i=1}^{s+1} \frac{g(x_i)}{\prod_{j \neq i}(x_i - x_j)} \tag{66}$$

Another identity that is proved by induction will be more important for us, namely

$$\Delta^s \; x^l(x_1, \ldots, x_{s+1}) = p_{l,s}(x_1, \ldots, x_{s+1}), \tag{67}$$

where $p_{l,s}(x_1, \ldots, x_{s+1})$ is 0 for $s > l$ and for $s \leq l$ is the sum of all degree $l - s$ monomials in $x_0, \ldots, x_s$ with unit coefficients,

$$p_{l,s}(x_1, \ldots, x_{s+1}) = \sum_{r_0 + \cdots + r_s = l - s} \; \prod_{j=1}^{s+1} x_j^{r_j}. \tag{68}$$

The divided differences form coefficients for the Newton interpolation formula. For all C^∞ functions $g\colon \mathbb{R} \to \mathbb{R}$ we have

$$\begin{aligned} g(x) = \Delta^0 g(x_1) + \Delta^1 g(x_1, x_2)(x - x_1) + \cdots \\ + \Delta^{k-1} g(x_1, \ldots, x_{k-2})(x - x_1) \cdots (x - x_{k-3}) \\ + \Delta^k g(x_1, \ldots, x_{k-1}, x)(x - x_1) \cdots (x - x_{k-2}) \end{aligned} \tag{69}$$

identically for all values of $x, x_1, \ldots, x_k$. All terms of this representation are polynomial in x except for the last one which we view as a remainder term. The sum of the polynomial terms is the degree $(k-1)$ *Newton interpolation polynomial* for g at

$\{x_s\}_{s=1}^k$. To obtain a degree $2k-1$ interpolation polynomial for g and its derivative at $\{x_s\}_{s=1}^k$, we simply use (69) with k replaced by $2k$ and the $2k$-tuple of points $\{x_{s \pmod k}\}_{s=1}^{2k}$. Similarly one can construct an interpolation polynomial for g and its derivatives up to any finite order.

All terms of this representation, except for the last one, are polynomial in x and their sum is the kth order *Newton Interpolation Polynomial* denoted by $\mathcal{P}_{k-1}(x, \mathbf{X}_k)$, where $\mathbf{X}_k = (x_1, \dots, x_k)$.

Now we can define entries of Figure 11 in the case $m = 0$. Let $\mathcal{DD}_k(I, \mathbb{R}) = \underbrace{I \times \cdots \times I}_{k \text{ times}} \times \mathbb{R}^k = (x_1, \dots, x_k; u_0, u_1, \dots, u_{k-1})$. It is called the *space of divided differences.* Then

$$
\begin{aligned}
&\mathcal{D}_k(f) : \underbrace{I \times \cdots \times I}_{k \text{ times}} \to \mathcal{DD}_k(I, \mathbb{R}), \\
&\mathcal{D}_k(f) : (x_1, \dots, x_k) \mapsto (x_1, \dots, x_k; u_0, \dots, u_{k-1}), \ u_\alpha = \Delta^\alpha f(x_1, \dots, x_{\alpha+1}) \\
&\pi_k \colon \mathcal{DD}_k(I, \mathbb{R}) \to \mathbb{R}^{2k},
\end{aligned}
\tag{70}
$$

where $\mathcal{D}_k(f)$ is smooth, provided that F is smooth, $\pi_n \colon \mathcal{DD}_k(I, \mathbb{R}) \to \mathbb{R}^{2k}$ is a Newton Interpolation Polynomial of the form (69), and

$$
\pi_k \circ \mathcal{D}_k(f) = (f, \dots, f) \colon I \times \cdots \times I \to \mathbb{R}^{2k},
\tag{71}
$$

where f and $\mathbb{R}$ are repeated k times.

4.2.2. *Language of divided differences and the Newton interpolation formula.* In this section we introduce the construction of the divided differences space $\mathcal{DD}_k(B^n, \mathbb{R}^N)$ and the corresponding map $\mathcal{D}_k(F)$ and the polynomial π_k presented on Figure 3.

Let $F \colon \mathbb{R}^n \to \mathbb{R}$ be a smooth function in n real variables $x_1, \dots, x_n$.

DEFINITION 21. The *first order divided difference* of F in the variable x_k is the function of $n+1$ variables $x_1, \dots, x_{k-1}, x_k', x_k'', \dots, x_n$ defined as

$$
\begin{aligned}
\Delta_{x_k} F(x_1, \dots, x_{k-1}, x_k', x_k'', \dots, x_n) \\
= \frac{F(x_1, \dots, x_{k-1}, x_k', \dots, x_n) - F(x_1, \dots, x_{k-1}, x_k'', \dots, x_n)}{x_k' - x_k''}
\end{aligned}
\tag{72}
$$

for $x_k' \neq x_k''$ and extended by its limit value as $\partial F(x_1, \dots, x_{k-1}, x_k', \dots, x_n)/\partial x_k$ for $x_k' = x_k'' = x_k$. Clearly, if F is a C^r function, then (e.g., by Hadamard lemma), $\Delta_{x_k} F$ is at least a C^{r-1}-smooth function of its arguments.

It turns out that iterating this construction is possible [**GY**], which leads to

DEFINITION 22. Let $\alpha = (\alpha_1, \dots, \alpha_n) \in \mathbb{Z}_+^n$ be a multiindex, let F be as above. Then $\Delta_x^\alpha F = \Delta_{x_1}^{\alpha_1} \cdots \Delta_{x_n}^{\alpha_n} F$ is called the mixed divided difference of order $|\alpha| = \alpha_1 + \cdots + \alpha_n$. This divided difference is a smooth function of $n + |\alpha|$ arguments subdivided into n groups of $\alpha_1 + 1, \dots, \alpha_n + 1$ variables, symmetric with respect to permutations of variables within the same groups.

As direct calculations show the operators Δ_{x_j} and Δ_{x_i} commute for $i \neq j$, and, therefore, we can use the multiindex notation for divided differences.

4.2.3. *The Newton interpolation formula (in multivariables)*. Let

$$X^1 = (x_1^1, \dots, x_N^1) \subset \mathbb{R}, \dots, X^n = (x_1^n, \dots, x_N^n) \subset \mathbb{R}$$

be subsets consisting of the same number of points, each X^j belonging to the corresponding jth coordinate axis of points in $\mathbb{R}^N$. Then, given a multiindex $\alpha \in \mathbb{Z}_+^N$ and a smooth function $F(x) = F(x^1, \dots, x^N)$ in N variables we can form the divided difference $\Delta_x^\alpha F(X^1, \dots, X^N)$.

In terms of the divided differences one can write the Newton interpolation polynomial as follows:

$$\mathcal{P}(t^1, \dots, t^N) = \sum_{0 \le \alpha_i \le n} \Delta_x^\alpha F(X^1, \dots, X^n) \prod_{i_1=1}^{\alpha_1} \cdots \prod_{i_n=1}^{\alpha_n} (t^1 - x_{i_1}^1) \cdots (t^n - x_{i_n}^n). \tag{73}$$

The polynomial $\mathcal{P}(t^1, \dots, t^n)$ has degree $\le kn$ in the variables $t = (t^1, \dots, t^n)$. The Newton interpolation formula implies that the difference $F(t) - \mathcal{P}(t^1, \dots, t^n)$ vanishes at all points of the Cartesian product grid $\mathbf{X} = X^1 \times \cdots \times X^n \subset \mathbb{R}^n$. Moreover, if for each $X^j = (x_1^j, \dots, x_n^j)$ we denote by $\operatorname{diag}^k(X^j)$ the set $(x_1^j, \dots, x_n^j)$ repeated $(m+1)$ times

$$(\underbrace{x_1^j, \dots, x_n^j}, \dots, \underbrace{x_1^j, \dots, x_n^j}) \quad (m \text{ times}), \tag{74}$$

then, to obtain interpolation of the mth jet of F, we replace each $X^j = (x_1^j, \dots, x_n^j)$ by $\operatorname{diag}^m(X^j)$. The degree of the interpolating polynomial will be $\le nk(m+1)$.

In the case of a multivariate function $F\colon B^n \to \mathbb{R}^N$ the interpolating polynomial $\mathcal{P}(t^1, \dots, t^n)$ becomes an N-dimensional vector and is interpolating by coordinate functions of $F = (F^1, \dots, F^N)$.

Definition 23. Let $\mathcal{DD}_k^m(B^n, \mathbb{R}^N)$ be the collection of all divided differences with m repetitions, $\{\Delta_x^\alpha F(\operatorname{diag}^m(X^1), \dots, \operatorname{diag}^m(X^n))\}_\alpha$, $\alpha_i \le (m+1)k$, $i = 1, \dots, n$. This is a linear space naturally equipped with the coordinates $\{x_i, u_\alpha : 0 \le i \le N, \alpha_i \le (m+1)\}$, where x_i (resp. u_α) are vectors from $\mathbb{R}^n$ (resp. $\mathbb{R}^N$). The dimension of this space is equal to $kn + N((m+1)k)^n$.

The map $\mathcal{D}_k^m F$ is defined by

$$\mathcal{D}_k^m F\colon (x_1, \dots, x_k) \to (x_1, \dots, x_k, \{u_\alpha\}_\alpha), \tag{75}$$

where $u_\alpha = \Delta_x^\alpha F$, $\forall i \ \alpha_i \le (m+1)n$.

The multivariate interpolation formula together with its derivatives in t_j evaluated at the points of the grid can be interpreted as a polynomial map restoring multijets from divided differences.

Newton Interpolation on $\mathbb{R}^m$ (abstract version). *The multivariate Newton interpolation formula* (73) *defines a polynomial interpolation map*

$$\pi_k^m\colon \mathcal{DD}_k^m(B^n, \mathbb{R}^N) \to \mathcal{J}^{m,k}(B^n, \mathbb{R}^N)$$

such that $\mathcal{J}^{m,k} f = \pi_k^m \circ \mathcal{D}_k^m F$. The degrees of the components of π_k^m do not exceed $(k+1)nN$.

In the next section we present an application of Newton interpolation polynomials and Figure 11 to an old problem in dynamical systems: the problem of rate of growth of the number of periodic points for generic diffeomorphisms (see, e.g., [**AM, S4**]).

4.3. Rate of growth of the number of periodic points for generic diffeomorphisms and Newton interpolation polynomials.

4.3.1. *Statement of the problem.* Let $\text{Diff}^r(M)$ be the space of C^r diffeomorphisms of a finite-dimensional smooth compact manifold M with the uniform C^r-topology, $\dim M \geq 2$, and let $f \in \text{Diff}^r(M)$. Consider the number of *isolated* periodic points of period n

$$P_n(f) = \#\{\text{isolated } x \in M : x = f^n(x)\}. \tag{76}$$

The main question of this paper is:

How quickly can $P_n(f)$ grow with n for a "generic" diffeomorphism f?

We put the word "generic" in brackets because as the reader will see the answer depends on the notion of genericity.

We call a diffeomorphism $f \in \text{Diff}^r(M)$ an *Artin – Mazur diffeomorphism* (or simply *A – M diffeomorphism)* if the number of isolated periodic orbits of f grows at most exponentially fast, i.e., for some number $C > 0$

$$P_n(f) \leq \exp(Cn) \quad \text{for all } n \in \mathbb{Z}_+. \tag{77}$$

Artin & Mazur [**AM**] proved the following

THEOREM 13. *For any* $0 \leq r \leq \infty$, *A – M diffeomorphisms form a dense set of diffeomorphisms in* $\text{Diff}^r(M)$ *with the uniform* C^r*-topology.*

In [**K3**] an elementary proof of the following extension of the Artin – Mazur result is given.

THEOREM 14. *For any* $0 \leq r < \infty$ *A – M diffeomorphisms with all periodic points hyperbolic are dense in* $\text{Diff}^r(M)$ *with the uniform* C^r*-topology.*

According to the standard terminology a set in $\text{Diff}^r(M)$ is called generic (or residual) if it contains a countable intersection of open dense sets and a property is called (Baire) generic if diffeomorphisms with that property form a residual set. It turns out the A – M property is not generic, as is shown in [**K4**]. Moreover:

THEOREM 15 ([**K4**]). *For any* $2 \leq r < \infty$ *there is an open set* $\mathcal{N} \subset \text{Diff}^r(M)$ *such that for any given sequence* $a = \{a_n\}_{n \in \mathbb{Z}_+}$ *there is a Baire generic set* $\mathcal{R}_a$ *in* $\mathcal{N}$ *depending on the sequence* a_n *with the property that if* $f \in \mathcal{R}_a$, *then for infinitely many* $n_i \in \mathbb{Z}_+$ *we have* $P_{n_i}(f) > a_{n_i}$.

Since any two residual sets have nonempty intersection Theorem 15 implies that A – M diffeomorphisms are not generic. The proof of this theorem is based on a result of Gonchenko – Shil′nikov – Turaev [**GST**]. Two slightly different detailed proofs of their result are given in [**K4, GTS**]. The proof in [**K4**] relies on a strategy outlined in [**GST**].

However, it seems unnatural that if you pick a diffeomorphism at random then it may have an arbitrarily fast growth of number of periodic points. Moreover, Baire generic sets in Euclidean spaces can have zero Lebesgue measure. Phenomena which are Baire generic but have a small probability are well known in dynamical systems, KAM theory, number theory, etc. (see [**O, HSY, K1**] for various examples).

This partially motivates the problem posed by Arnol′d [**A**]: *Prove that "with probability one" f is an A – M diffeomorphism.* Arnol′d suggested the following interpretation of "with probability one": *for a (Baire) generic finite parameter* *family of diffeomorphisms* $\{f_\varepsilon\}$, *for Lebesgue almost every* ε *we have that* f_ε *is* $A - M$. (cf. [**K1**]). As Theorem 15 shows, a result on the genericity of the set of A – M diffeomorphisms based on topology (Baire) is likely to be extremely subtle, if possible at all.[5] We use instead a notion of "probability one" based on prevalence [**HSY, K1**], which is independent of Baire genericity. We also are able to state the result in the form Arnol′d suggested for generic families using this measure-theoretic notion of genericity. The main result in this direction is a partial solution to Arnol′d's problem. It says that *for a prevalent diffeomorphism* $f \in \mathrm{Diff}^r(M)$, $r > 1$, *and any* $\delta > 0$ *there exists* $C = C(\delta) > 0$ *such that for all* $n \in \mathbb{Z}_+$

$$P_n(f) \le \exp(Cn^{1+\delta}) \tag{78}$$

This theorem is announced in [**KH**]. A major part of the proof is worked out in [**K5**]. We omit the precise statement, which requires an additional discussion.

4.4. Dynamical usage of Newton interpolation polynomials.

4.4.1. *Perturbation of recurrent trajectories by Newton interpolation polynomials.* Let us start with several remarks which were the starting point of this paper. In order to keep notations and formulas simple we consider the 1-dimensional maps, but the reader should always have in mind that our consideration is designed for multidimensional diffeomorphisms.

Consider a map $f\colon I \hookrightarrow I$ of the interval $I = [-1, 1]$. Recall that a trajectory $\{x_k\}_{k\in\mathbb{Z}}$ of f is called *recurrent* if it returns arbitrarily close to its initial position — that is, for all $\delta > 0$ we have $|x_0 - x_n| < \delta$ for some $n > 0$. A very basic question of closing lemma type is how much one should perturb f to create a periodic point x_0. Let us give a "baby" answer

Baby Closing Lemma. *Let* $\{x_k = f^k(x_0)\}_{k=0}^n$ *be a trajectory of length* $n+1$ *of a map* $f\colon I \hookrightarrow I$. *Let* $u = (x_n - x_0)/\prod_{k=0}^{n-2}(x_{n-1} - x_k)$. *Then* x_0 *is a periodic point of period* n *of the map*

$$\tilde f(x) = f(x) + u \prod_{k=0}^{n-2} (x - x_k) \tag{79}$$

Of course $\tilde f$ is close to f only if u is sufficiently small, meaning that $|x_0 - x_n|$ is small compared to $\prod_{k=0}^{n-2}(x_{n-1} - x_k)$. However, this product is likely to contain small factors for a recurrent trajectory. In general, it is difficult to control the effect of perturbations for recurrent trajectories. The simple reason is because *one cannot perturb* f *at two nearby points independently.*

It is important for the proof in [**K5**] to control the derivative of f along periodic orbits. If for some $x \in I$, $\gamma > 0$ and some positive integer n we have $f^n(x) = x$ and $|(f^n)'(x) - 1| > \gamma$, then this implies that the interval around x of size $\|f\|_{C^1}^{-n}\gamma$ is free from periodic points of the same period (see [**KH**, Proposition 1.1]). The quantity γ is called *hyperbolicity* and x is called (n, γ)-hyperbolic. This quantity was introduced by Gromov [**G1**] and Yomdin [**Y2**]. If one can estimate hyperbolicity for all points of period n from below, then one can estimate the number of periodic points of period n. The upper bound (78) is obtained by proving a lower bound on

[5] For example, using techniques from [**GTS, K4**] one can prove that for a Baire generic finite-parameter family $\{f_\varepsilon\}$ and a Baire generic parameter value ε the corresponding diffeomorphism f_ε is not A – M. Unfortunately, how to estimate from below the measure of non-A – M diffeomorphisms in a Baire generic finite-parameter family is so far an unreachable question.

the rate of decay of hyperbolicity with period for prevalent diffeomorphisms. This is the reason the proof needs to control the derivative along trajectories.

The closing lemma above also gives an idea of how much we must change the parameter u to make a point x_0 that is (n,γ)-periodic not be (n,γ)-periodic for a given $\gamma > 0$, which as we described above is one way to make a map that is "bad" for the initial condition x_0 become "good." To make use of our other alternative we must determine how much we need to perturb a map f to make a given x_0 be (n,γ)-hyperbolic for some $\gamma > 0$.

PERTURBATION OF HYPERBOLICITY. *Let $\{x_k = f^k(x_0)\}_{k=0}^{n-1}$ be a trajectory of length n of a C^1 map $f : I \to I$. Then for the map*

$$f_v(x) = f(x) + v(x - x_{n-1}) \prod_{k=0}^{n-2} (x - x_k)^2 \tag{80}$$

such that $v \in \mathbb{R}$ and

$$\big||(f_v^n)'(x_0)| - 1\big| = \left|\left|\prod_{k=0}^{n-1} f'(x_k) + v \prod_{k=0}^{n-2} (x_{n-1} - x_k)^2 \prod_{k=0}^{n-2} f'(x_k)\right| - 1\right| > \gamma \tag{81}$$

we have that x_0 is an (n,γ)-hyperbolic point of f_v.

Once again we can see that the product of distances $\prod_{k=0}^{n-2} |x_{n-1} - x_k|$ along the trajectory is an important quantitative characteristic of how much freedom we have to perturb.

The perturbations (79) and (80) are reminiscent of Newton interpolation polynomials. Let us put these formulas into a general setting using singularity theory.

4.4.2. *Distance to the diagonal in the multijet space.* Consider the $2n$-parameter family of a perturbation of a map $f\colon I \hookrightarrow I$ by polynomials of degree $2n-1$

$$f_\varepsilon(x) = f(x) + \sum_{k=0}^{2n-1} \varepsilon_k x^k. \tag{82}$$

Define a map

$$\begin{gathered} \mathcal{J}_n^1 f\colon \underbrace{I \times \cdots \times I}_{n \text{ times}} \times \mathbb{R}^{2n} \to \underbrace{I \times \cdots \times I}_{n \text{ times}} \times \underbrace{(I \times \mathbb{R}) \times \cdots \times (I \times \mathbb{R})}_{n \text{ times}} \\ \mathcal{J}^{1,n} f(x_0, \dots, x_{n-1}, \varepsilon) = \big(x_0, \dots, x_{n-1}, f_\varepsilon(x_0), f'_\varepsilon(x_0), \dots, f_\varepsilon(x_{n-1}), f'_\varepsilon(x_{n-1})\big). \end{gathered} \tag{83}$$

This map is called the *n-tuple* 1*-jet map.* The 1-jet of a function means that we take into account not only the image of a point, but also its derivative. The 1-jet of a function is usually denoted by $j^1 f_\varepsilon(x) = (x, f_\varepsilon(x), f'_\varepsilon(x))$. The space of 1-jets of functions on the interval I is denoted by $\mathcal{J}^1(I,\mathbb{R})$. The product of n copies of $\mathcal{J}^1(I,\mathbb{R})$ is *multijet space* and is denoted by

$$\mathcal{J}^{1,n}(I,\mathbb{R}) = \underbrace{\mathcal{J}^1(I,\mathbb{R}) \times \cdots \times \mathcal{J}^1(I,\mathbb{R})}_{n \text{ times}}. \tag{84}$$

We need to include derivatives into our consideration, because we are interested in hyperbolicity (property of derivatives) of periodic points. The set of points

$$\Delta_n(I) = \Big\{\{x_0, \dots, x_{n-1}\} \times \mathbb{R}^{2n} \subset \underbrace{I \times \cdots \times I}_{n \text{ times}} \times \mathbb{R}^{2n} : \exists (i,j) \text{ s.t. } i \neq j,\ x_i = x_j\Big\} \tag{85}$$

is called the *diagonal* in the space of multijets. In singularity theory the space of multijets is defined outside of the diagonal $\Delta_n(I)$ and is usually denoted by $\mathcal{J}^{1,n}(I,\mathbb{R}) = \mathcal{J}^{1,n}(I,\mathbb{R}) \setminus \Delta_n(I)$ (see [**GG**]).

It is easy to see that a *recurrent trajectory* $\{x_k\}_{k\in\mathbb{Z}_+}$ *is located in a neighborhood of the diagonal* $\Delta_n(I)$ *in the space of multijets for a sufficiently large* n. If $\{x_k\}_{k=0}^{n-1}$ is a part of a recurrent trajectory of length n, then the product of distances along the trajectory

$$\prod_{k=0}^{n-2} |x_{n-1} - x_k| \tag{86}$$

measures how close $\{x_k\}_{k=0}^{n-1}$ is to the diagonal $\Delta_n(I)$, or how independently one can perturb points of a trajectory. One can also say that (86) is a quantitative characteristic of how recurrent a trajectory of length n is. Introduction of this *product of distances along a trajectory* is a new central point of the method.

4.4.3. *Newton interpolation and blow-up along the diagonal in multijet space.* Now look at Grigoriev – Yakovenko's construction [**GY**] in the 1-dimensional case with more details. This construction puts the "closing lemma" and "perturbation of hyperbolicity" statements above into a general framework.

Again consider the $2n$-parameter family (82) of perturbations of a C^1 map $f\colon I \to I$ by polynomials of degree $2n-1$. Our goal now is to describe how such perturbations affect the n-tuple 1-jet of f, and since the operator $j^{1,n}$ is linear in f, for the time being we consider only the perturbations ϕ_ε and their n-tuple 1-jets. For each n-tuple $\{x_k\}_{k=0}^{n-1}$ there is a natural transformation $\mathcal{J}^{1,n}\colon I^n\times\mathbb{R}^{2n} \to \mathcal{J}^{1,n}(I,\mathbb{R})$ from ε-coordinates to jet-coordinates, given by

$$\mathcal{J}^{1,n}(x_0,\dots,x_{n-1},\varepsilon) = j^{1,n}\phi_\varepsilon(x_0,\dots,x_{n-1}). \tag{87}$$

Instead of working directly with the transformation $\mathcal{J}^{1,n}$, we introduce intermediate u-coordinates based on Newton interpolation polynomials. The relation between ε-coordinates and u-coordinates is given implicitly by

$$\phi_\varepsilon(x) = \sum_{k=0}^{2n-1} \varepsilon_k x^k = \sum_{k=0}^{2n-1} u_k \prod_{j=0}^{k-1} (x - x_{j(\mathrm{mod}\, n)}). \tag{88}$$

Based on this identity, we can define functions $\mathcal{D}_n^1\colon I^n \times \mathbb{R}^{2n} \to I^n \times \mathbb{R}^{2n}$ and $\pi_n^1\colon I^n \times \mathbb{R}^{2n} \to \mathcal{J}^{1,n}(I,\mathbb{R})$ so that $\mathcal{J}^{1,n} = \pi_n^1 \circ \mathcal{D}_n^1$, or in other words the diagram in Figure 11 commutes. This definition coincides with the one we gave before. We will show later that $\mathcal{D}_n^1$ is invertible, while π_n^1 is invertible away from the diagonal $\Delta_n(I)$ and defines a blow-up along it in the space of multijets $\mathcal{J}^{1,n}(I,\mathbb{R})$. Consider Figure 11 for $m = n = N = 1$.

Recall that the intermediate space, denoted by $\mathcal{DD}_n^1(I,\mathbb{R})$, is called the *space of divided differences* and consists of n-tuples of points $\{x_k\}_{k=0}^{n-1}$ and $2n$ real coefficients $\{u_k\}_{k=0}^{2n-1}$. Here are explicit coordinate-by-coordinate formulas defining

$\pi_n^1 \colon \mathcal{DD}_n^1(I,\mathbb{R}) \to \mathcal{J}^{1,n}(I,\mathbb{R})$.

$$
\begin{aligned}
&\phi_\varepsilon(x_0) = u_0,\\
&\phi_\varepsilon(x_1) = u_0 + u_1(x_1 - x_0),\\
&\phi_\varepsilon(x_2) = u_0 + u_1(x_2 - x_0) + u_2(x_2 - x_0)(x_2 - x_1),\\
&\qquad\vdots\\
&\phi_\varepsilon(x_{n-1}) = u_0 + u_1(x_{n-1} - x_0) + \cdots + u_{n-1}(x_{n-1} - x_0)\cdots(x_{n-1} - x_{n-2}),
\end{aligned}
\tag{89}
$$

$$
\begin{aligned}
\phi'_\varepsilon(x_0) &= \frac{\partial}{\partial x}\Bigg(\sum_{k=0}^{2n-1} u_k \prod_{j=0}^{k-1}\big(x - x_{j(\operatorname{mod} n)}\big)\Bigg)\Bigg|_{x=x_0},\\
&\ \ \vdots\\
\phi'_\varepsilon(x_{n-1}) &= \frac{\partial}{\partial x}\Bigg(\sum_{k=0}^{2n-1} u_k \prod_{j=0}^{k-1}\big(x - x_{j(\operatorname{mod} n)}\big)\Bigg)\Bigg|_{x=x_{n-1}},
\end{aligned}
\tag{90}
$$

These formulas are very useful for dynamics. For a given base map f and initial point x_0, the image $f_\varepsilon(x_0) = f(x_0) + \phi_\varepsilon(x_0)$ of x_0 depends only on u_0. Furthermore the image can be set to any desired point by choosing u_0 appropriately — we say then that it depends nontrivially on u_0. If x_0, x_1, and u_0 are fixed, the image $f_\varepsilon(x_1)$ of x_1 depends only on u_1, and as long as $x_0 \neq x_1$ it depends nontrivially on u_1. More generally for $0 \le k \le n-1$, if pairwise distinct points $\{x_j\}_{j=0}^{k}$ and coefficients $\{u_j\}_{j=0}^{k-1}$ are fixed, then the image $f_\varepsilon(x_k)$ of x_k depends only and nontrivially on u_k.

Suppose now that an n-tuple of points $\{x_j\}_{j=0}^{n}$ not on the diagonal $\Delta_n(I)$ and Newton coefficients $\{u_j\}_{j=0}^{n-1}$ are fixed. Then derivative $f'_\varepsilon(x_0)$ at x_0 depends only and nontrivially on u_n. Likewise for $0 \le k \le n-1$, if distinct points $\{x_j\}_{j=0}^{n}$ and Newton coefficients $\{u_j\}_{j=0}^{n+k-1}$ are fixed, then the derivative $f'_\varepsilon(x_k)$ at x_k depends only and nontrivially on u_{n+k}.

As Figure 12 illustrates, these considerations show that for any map f and any desired trajectory of distinct points with any given derivatives along it, one can choose Newton coefficients $\{u_k\}_{k=0}^{2n-1}$ and explicitly construct a map $f_\varepsilon = f + \phi_\varepsilon$ with such a trajectory. Thus we have shown that π_n^1 is invertible away from the diagonal $\Delta_n(I)$ and defines a blow-up along it in the space of multijets $\mathcal{J}^{1,n}(I,\mathbb{R})$.

The function $\mathcal{D}^{1,n} \colon I^n \times \mathbb{R}^{2n} \to \mathcal{DD}^{1,n}(I,\mathbb{R})$ was explicitly defined using so-called divided differences above.

Recall that $\mathcal{D}_n^1$ was defined implicitly by (88). We have described how to use divided differences to construct a degree $2n-1$ interpolating polynomial of the form on the right-hand side of (88) for an arbitrary C^∞ function g. Our interest then is in the case $g = \phi_\varepsilon$, which as a degree $2n-1$ polynomial itself will have no remainder term and coincide exactly with the interpolating polynomial. Thus $\mathcal{D}^{1,n}$ is given coordinate-by-coordinate by

$$
\begin{aligned}
u_m &= \Delta^m\Bigg(\sum_{k=0}^{2n-1} \varepsilon_k x^k\Bigg)\big(x_0, \ldots, x_{m(\operatorname{mod} n)}\big)\\
&= \varepsilon_m + \sum_{k=m+1}^{2n-1} \varepsilon_k p_{k,m}\big(x_0, \ldots, x_{m(\operatorname{mod} n)}\big)
\end{aligned}
\tag{91}
$$

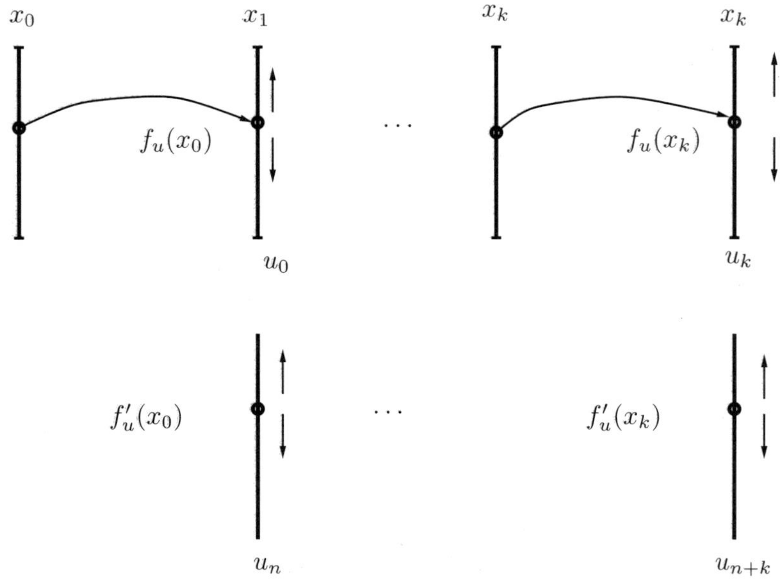

FIGURE 12. Newton coefficients and their action

for $m = 0, \dots, 2n-1$. We call the transformation given by (91) the *Newton map*. Notice that for fixed $\{x_k\}_{k=0}^{2n-1}$, the Newton map is linear and given by an upper triangular matrix with units on the diagonal. Hence it is Lebesgue volume-preserving and invertible, whether or not $\{x_k\}_{k=0}^{2n-1}$ lies on the diagonal $\Delta_n(I)$.

We call the basis of monomials

$$\prod_{j=0}^{k}(x - x_{j(\operatorname{mod} n)}) \quad \text{for } k = 0, \dots, 2n-1 \tag{92}$$

in the space of polynomials of degree $2n-1$ the *Newton basis* defined by the n-tuple $\{x_k\}_{k=0}^{n-1}$. The Newton map and the Newton basis, and their analogues in dimension N, are useful tools for perturbing trajectories and proving (78).

Acknowledgments. In this lecture I have used fragments of the announcement [**KH**]. The good presentation of this announcement is entirely due to my coauthor Brian Hunt. Needless to say that numerous communications with him and also John Mather were very important for me.

References

[AA] D. V. Anosov and V. I. Arnol'd (eds.), *Dynamical systems.* I, Encyclopaedia Math. Sci., vol. 1, Springer-Verlag, Berlin, 1988.

[A] V. Arnol'd, *Arnold problems*, with preface by M. B. Sevryuk and V. B. Filippov, Izdatel'stvo FAZIS, Moscow, 2000 (Russian).

[AGV] V. I. Arnol'd, S. M. Guseĭn-Zade, and A. N. Varchenko, *Singularities of differentiable maps.* Vol. I: *The classification of critical points, caustics and wave fronts*, Monogr. Math., vol. 82, Birkhäuser, Boston, MA, 1985.

[AM] M. Artin and B. Mazur, *On periodic points*, Ann. of Math. (2) **81** (1965), 82–99.

[AVI+] V. I. Arnol'd, M. I. Vishik, Yu. S. Ilyashenko, A. S. Kalashnikov, V. A. Kondrat'ev, S. N. Kruzhkov, E. M. Landis, V. M. Millionshchikov, O. A. Oleĭnik, A. F. Filippov, and M. A. Shubin, *Some unsolved problems in the theory of differential equations and mathematical physics*, Uspekhi Mat. Nauk **44** (1989), no. 4(268), 191–202 (Russian); English transl., Russian Math. Surveys **44** (1989), no. 4, 157–171.

[B] I. Bendixson, *Sur les courbes définies par des équations différentielles*, Acta Math. **24** (1901), 1–88.

[BS] I. S Berezin and N. P. Shidkov, *Computing methods.* Vol. I, Pergamon, Oxford, 1965.

[BCR] J. Bochnak, M. Coste, and M.-F. Roy, *Real algebraic geometry*, Ergeb. Math. Grenzgeb. (3), vol. 36, Springer-Verlag, Berlin, 1998.

[B1] A. A. Bolibrukh, *Linear differential equations, Fuchsian inequalities and multiplicities of zeros*, in this volume.

[B2] A. Buium, *Finiteness results in differential algebraic geometry and Diophantine geometry*, in this volume.

[DR] Z. Denkowska and R. Roussarie, *A method of desingularization for analytic two-dimensional vector field families*, Bol. Soc. Brasil. Mat. (N.S.) **22** (1991), no. 1, 93–126.

[DW] Z. Denkowska and K. Wachta, *Une construction de la stratification sous-analytique avec la condition* (w), Bull. Polish Acad. Sci. Math. **35** (1987), no. 7-8, 401–405.

[vdD] L. van den Dries, *o-minimal structures and real analytic geometry*, Current Developments in Mathematics (Cambridge, MA, 1998), Int. Press, Somerville, MA, 1999, pp. 105–152.

[vdDS] L. van den Dries and P. Speisseger, *o-minimal preparation theorems*, Quad. Mat., (to appear).

[D1] H. Dulac, *Sur les cycles limites*, Bull. Soc. Math. France **51** (1923), 45–188.

[D2] F. Dumortier, *Singularities of vector fields on the plane*, J. Differential Equations **23** (1977), no. 1, 53–106.

[DMR] F. Dumortier, M. El Morsalani, and C. Rousseau, *Hilbert's 16th problem for quadratic systems and cyclicity of elementary graphics*, Nonlinearity **9** (1996), no. 5, 1209–1261.

[DRR] F. Dumortier, R. Roussarie, and C. Rousseau, *Hilbert's 16th problem for quadratic vector fields*, J. Differential Equations **110** (1994), no. 1, 86–133.

[É] J. Écalle, *Introduction aux fonctions analysables et preuve consrustive de la conjecture de Dulac*, Actualités Math., Herman, Paris, 1992.

[FP] J.-P. Françoise and C. C. Pugh, *Keeping track of limit cycles*, J. Differential Equations **65** (1986), no. 2, 139–157.

[GAK] Gabrielov A. and A. Khovanskiĭ, *Multiplicity of a Noetherian intersection*, Geometry of Differential Equations, Amer. Math. Soc. Transl. Ser. 2, vol. 186, Amer. Math. Soc., Providence, RI, 1998, pp. 119–131.

[G] L. Gavrilov, *The infinitesimal 16th Hilbert problem in the quadratic case*, Invent. Math. **143** (2001), no. 3, 449–497.

[GWdPL] C. G. Gibson, K. Wirthmüller, A. A du Plessis, and E. J. N. Loojenga, *Topological stability of smooth mappings*, Lecture Notes in Math., vol. 552, Springer-Verlag, Berlin–New York, 1976.

[GG] M. Golubitsky and V. Guillemin, *Stable mappings and their singularities*, Grad. Texts in Math., vol. 14, Springer-Verlag, New York–Heidelberg, 1973.

[GST] S. V. Gonchenko, L. P. Shil′nikov, and D. V. Tuvaev, *On models with nonrough Poincaré homoclinic curves*, Phys. D **62** (1993), no. 1-4, 1–14.

[GTS] S. V. Gonchenko, D. V. Tuvaev, and L. P. Shil′nikov, *Homoclinic tangencies of an arbitrary order in Newhouse domains*, Dynamical Systems (Moscow, 1998), Itogi Nauki Tekh. Ser. Sovrem. Mat. Prilozh. Temat. Obz., vol. 67, Vseross. Inst. Nauchn. i Tekhn. Inform. (VINITI), Moscow, 1999, pp. 69–128 (Russian).

[GM] M. Goresky and R. MacPherson, *Stratified Morse theory*, Ergeb. Math. Grenzgeb. (3), vol. 14, Springer-Verlag, Berlin, 1988.

[GY] A. Grigoriev and S. Yakovenko, *Topology of generic multijet preimages and blow-up via Newton interpolation*, J. Differential Equations **150** (1998), no. 2, 349–362.

[G1] M. Gromov, *On entropy of holomorphic maps* (1985), (preprint).

[G2] T. Grozovskiĭ, *Bifurcations of "apple" and "half-apple" polycycles in generic two-parameter families*, Differ. Uravn. **32** (1996), no. 4, 458–469; English transl., Differential Equations **32** (1996), no. 4, 459–470.

[GH] J. Guckenheimer and P. Holmes, *Nonlinear oscillations, dynamical systems, and bifurcations of vector fields*, Appl. Math. Sci., vol. 42, Springer-Verlag, New York, 1983.

[H] H. Hironaka, *Subanalytic sets*, Number Theory, Algebraic Geometry and Commutative Algebra, in Honor of Yasuo Akizuki, Kinokuniya, Tokyo, 1973, pp. 453–493.

[HSY] B. R. Hunt, T. Sauer, and J. A. Yorke, *Prevalence: a translation-invariant 'almost every' on infinite-dimensional spaces.*, Bull. Amer. Math. Soc. (N.S.) **27** (1992), no. 2, 217–238; addendum, Bull. Amer. Math. Soc. (N.S.) **28** (1993), no. 2, 306–307.

[I1] Yu. S. Ilyashenko, *The multiplicity of limit cycles that arise in the perturbation of a Hamiltonian equation of the class* $\omega' = P_2/Q_1$ *in a real and complex domain*, Trudy Sem. Petrovsk. **3** (1978), 49–60 (Russian); English transl., 191–202.

[I2] ———, *Dulac's memoir "On limit cycles" and related questions of the local theory of differential equations*, Uspekhi Mat. Nauk **40** (1985), no. 6(246), 41–78 (Russian); English transl., Russian Math. Surveys **40** (1985), no. 6, 1–50.

[I3] ———, *Finiteness theorem for limit cycles*, Transl. Math. Monogr., vol. 94, Amer. Math. Soc., Providence, RI, 1991.

[I4] ———, *Normal forms for local families and nonlocal bifurcations*, Complex Analytic Methods in Dynamical Systems (Rio de Janeiro, 1992), Astérisque, vol. 222, Soc. Math. France, Paris, 1994, pp. 233–258.

[IK] Yu. S. Ilyashenko and V. Yu. Kaloshin, *Bifurcation of planar and spatial polycycles: Arnold's program and its development*, The Arnoldfest (Toronto, 1997), Fields Inst. Commun., vol. 24, Amer. Math. Soc., Providence, RI, 1999, pp. 241–271.

[IL] Yu. S. Ilyashenko and W. Li, *Nonlocal bifurcations*, Math. Surveys Monogr., vol. 66, Amer. Math. Soc., Providence, RI, 1999.

[IY1] Yu. S. Ilyashenko and S. Yakovenko, *Finitely smooth normal forms of local families of diffeomorphisms and vector fields*, Uspekhi Mat. Nauk **46** (1991), no. 1(277), 3–39 (Russian); English transl., Russian Math. Surveys **46** (1991), no. 1, 1–43.

[IY2] ———, *Finite cyclicity of elementary polycycles in generic families*, Concerning the Hilbert 16th Problem, Amer. Math. Soc. Transl. Ser. 2, vol. 165, Amer. Math. Soc., Providence, RI, 1995.

[J] N. Jacobson, *Basic algebra*. I, W. H. Freeman and Co., San Francisco, CA, 1974.

[K1] V. Yu. Kaloshin, *Some prevalent properties of smooth dynamical systems*, Tr. Mat. Inst. Steklova **213** (1997), 123–151 (Russian); English transl., Proc. Steklov Inst. Math. **1996**, no. 2(213), 115–140.

[K2] ———, *Prevalence in spaces of finitely smooth mappings*, Funktsional. Anal. i Prilozhen. **31** (1997), no. 2, 27–33 (Russian); English transl., Funct. Anal. Appl. **31** (1997), no. 2, 95–99.

[K3] ———, *An extension of the Artin–Mazur theorem*, Ann of Math. (2) **50** (1999), no. 2, 729–741.

[K4] ———, *Generic diffeomorphisms with superexponential growth of number of periodic points*, Comm. Math. Physics **211** (2000), no. 1, 253–271.

[K5] ———, *Stretched exponential estimate on the rate of growth of the number of periodic points for prevalent diffeomorphisms*, Princeton University, 2001.

[K6] ———, *The Hilbert–Arnol'd problem and estimates on cyclicity of elementary polycycles in the plane and in space*, Funktsional. Anal. i Prilozhen. **35** (2001), no. 2, 78–81 (Russian); English transl., Functional Anal. Appl. **35** (2001), no. 2, 146–147.

[K7] ———, *A geometric proof of existence of Whitney's stratifications*, Moscow Math. J., (to appear).

[K8] ———, *Bifurcations of spatial polycycles*, (in preparation) (Russian).

[KH] V. Yu. Kaloshin and B. R. Hunt, *A stretched exponential estimate on the rate of growth of the number of periodic points for prevalent diffeomorphisms*. I, Electron. Res. Announc. Amer. Math. Soc. **7** (2001), 17–27; II, 28–36.

[K1] A. G. Khovanskiĭ, *Real analytic manifolds with the property of finiteness and complex abelian integrals*, Funktsional. Anal. i Prilozhen. **18** (1984), no. 2, 40–50 (Russian); English transl., Functional Anal Appl. **18** (1984), no. 2, 119–127.

[K2] ———, *Fewnomials*, Transl. Math. Monogr., vol. 88, Amer. Math. Soc., Providence, RI, 1991.

[K3] O. Kleban, *Order of the topologically sufficient jet of a smooth vector field on the real plane at a singular point of finite multiplicity*, Concerning the Hilbert 16th Problem, Amer. Math. Soc. Transl. Ser. 2, vol. 165, Amer. Math. Soc., Providence, RI, 1995.

[KS] A. Kotova and V. Stanzo, *On few-parameter generic families of vector fields on the two-dimensional sphere*, Concerning the Hilbert 16th Problem, Amer. Math. Soc. Transl. Ser. 2, vol. 165, Amer. Math. Soc., Providence, RI, 1995, pp. 155 – 201.

[K] T.-C. Kuo, *The ratio test for analytic Whitney stratifications*, Proceedings of Liverpool Singularities-Symposium. I (1969/70), Lecture Notes in Math., vol. 192, Springer, Berlin, 1971, pp. 141 – 149.

[Ł1] S. Łojasiewicz, *Ensembles semi-analytiques*, Institut de Hautes Études Scientifiques, 1965, (lecture notes).

[Ł2] ———, *Sur le géometrie semi- et sous-analytique*, Ann. Inst. Fourier (Grenoble) **43** (1993), no. 5, 1575 – 1595.

[ŁSW] S. Łojasiewicz, J. Stasica, and K. Wachta, *Stratifications sous-analytiques. Condition de Verdier*, Bull. Polish Acad. Sci. Math. **34** (1986), no. 9-10, 531 – 539.

[M1] P. Mardešić, *An explicit bound for the multiplicity of zeros of generic Abelian integrals*, Nonlinearity **4** (1991), no. 3, 845 – 852.

[M2] J. Mather, *Notes on topological stability*, Harvard University, 1970.

[M3] J. W. Milnor, *Singular points of complex hypersurfaces*, Ann. of Math. Stud., vol. 61, Princeton Univ. Press, Princeton, NJ, 1968.

[M4] ———, *Topology from the differentiable viewpoint*, Princeton Landmarks Math., Princeton Univ. Press, Princeton, NJ, 1997.

[MR] R. Moussu and C. Roche, *Théorie de Hovanskiĭ et problème de Dulac*, Invent. Math. **105** (1991), no. 2, 431 – 441.

[M] D. Mumford, *Algebraic geometry. I: Complex projective varieties*, Grundlehren Math. Wiss., vol. 221, Springer-Verlag, Berlin – New York, 1976.

[NDY] Novikov D. and S. Yakovenko, *Simple exponential estimate for the number of zeroes of complete Abelian integrals*, Ann. Inst. Fourier (Grenoble) **45** (1995), no. 4, 897 – 927.

[O] J. Oxtoby, *Measure and category: A survey of the analogies between topological and measure spaces*, Grad. Texts in Math., vol. 2, Springer-Verlag, New York – Berlin, 1980.

[P1] W. Pawłucki, *Le théorème de Puiseux pour une application sous-analytique*, Polish Acad. Sci. Math. **32** (1984), no. 9-10, 555 – 560.

[P2] G. S. Petrov, *Nonoscillation of elliptic integrals*, Funktsional. Anal. i Prilozhen. **24** (1990), no. 3, 45 – 50 (Russian); English transl., Funct. Anal. Appl. **24** (1990), no. 3, 205 – 210.

[dPW] A. du Plessis and T. Wall, *The geometry of topological stability*, London Math. Soc. Monogr. (N.S.), vol. 9, Oxford Univ. Press, New York, 1995.

[R1] R. Roussarie, *A note on finite cyclicity property and Hilbert's 16th problem*, Dynamical Systems (Valparaíso, 1986) (R. Bamón, R. Labarca, and J. Palis Jr., eds.), Lecture Notes in Math., vol. 1331, Springer-Verlag, Berlin, 1988, pp. 161 – 168.

[R2] ———, *Bifurcations of planar vector fields and Hilbert's sixteenth problem*, Progr. Math., vol. 164, Birkhäuser, Basel, 1998.

[S1] D. Schlomiuk (ed.), NATO Adv. Sci. Inst. Ser. C Math. Phys. Sci., vol. 408, Kluwer, Dordrecht, 1993.

[S2] A. Seidenberg, *Reduction of singularities of the differential equations* $A\,dy = B\,dx$, Amer. J. Math. **90** (1968), 248 – 269.

[S3] L. P. Shil′nikov, *A case of the existence of a denumerable set of periodic motions*, Dokl. Akad. Nauk SSSR **160** (1965), 558 – 561 (Russian).

[S4] S. Smale, *Differentiable dynamical systems*, Bull. Amer. Math. Soc. **73** (1967), 747 – 817.

[T1] F. Takens, *Unfoldings of certain singularities of vector fields: generalized Hopf bifurcations*, J. Differential Equations **14** (1973), 476 – 493.

[T2] R. Thom, *Ensembles et morphismes stratifiés*, Bull. Amer. Math. Soc. **75** (1969), 240 – 284.

[T3] ———, *Propriétés différentielles locales des ensembles analytiques (d'après H. Whitney)*, Exp. No. 281, Séminaire Bourbaki, Vol. 9, Soc. Math. France, Paris, 1995, pp. 69 – 80.

[T4] S. Trifonov, *Desingularization in families of analytic differential equations*, Concerning the Hilbert 16th Problem, Amer. Math. Soc. Transl. Ser. 2, vol. 165, Amer. Math. Soc., Providence, RI, 1995, pp. 97 – 129.

[V] A. Varchenko, *Estimation of the number of zeros of an abelian integral depending on a parameter, and limit cycles*, Funktsional. Anal. i Prilozhen. **18** (1984), no. 2, 14–25 (Russian); English transl., Functional Anal. Appl. **18** (1984), no. 2, 98–108.

[W1] T. C. Wall, *Regular stratifications*, Dynamical Systems (Warwick, 1974), Lecture Notes in Math., vol. 468, Springer-Verlag, Berlin–New York, 1975, pp. 332–344.

[W2] H. Whitney, *Tangents to an analytic variety*, Ann. of Math. (2) **81** (1965), 496–549.

[Y1] S. Yakovenko, *Quantitative theory of ordinary differential equations and tangential Hilbert 16th problem*, in this volume.

[Y2] Y. Yomdin, *A quantitative version of the Kupka–Smale theorem*, Ergodic Theory Dynam. Systems **5** (1985), no. 3, 449–472.

CHAPTER 5

Finiteness Results in Differential Algebraic Geometry and Diophantine Geometry

Alexandru Buium

The main purpose of these lectures is to illustrate the interaction between "differential algebraic geometry" and Diophantine geometry over function fields, following our papers [**B2, B3, B4, B5, B8**]. Differential algebraic geometry is an extension of usual algebraic geometry in which algebraic equations are replaced by algebraic differential equations; its "commutative algebra counterpart" is the Ritt–Kolchin "differential algebra" [**K**]. It turned out (cf. [**B2, B3, B4, B5, B8**]) that some of the fundamental (finiteness) conjectures in Diophantine geometry have analogues in differential algebraic geometry which in their turn imply the geometric (i.e., function field) versions of the Diophantine conjectures. The present lectures intend to be a (very sketchy) introduction to this circle of ideas. For a more systematic (although far from comprehensive) introduction we refer to our book [**B5**]; for the full story we refer to our papers [**B2, B3, B4, B5, B8**]. The circle of ideas presented in these lectures has a purely arithmetic (p-adic) counterpart which will not be discussed here at all; for this counterpart we refer to our papers [**B6, B7**].

The following notes are based on a series of lectures given by the author at the workshop on "Asymptotic series, differential algebra, and finiteness problems in nonlinear dynamical systems" organized by Centre de recherches mathématiques at Université de Montréal in June–July 2000. They differ from the oral exposition in that they provide less background but, at the same time, they cover more material. In particular the circle of ideas around the *abc* conjecture, which we included in the notes, was not discussed during the workshop; for our discussion of *abc* here we have borrowed from our lectures on this topic held at the Arizona Winter School, in Tucson, 1998. The format of these notes is also slightly different from that of the oral presentation; we rearranged the material of the five one hour lectures into three denser lectures. The author would like to thank the organizers of the workshop for their hospitality and for the stimulating atmosphere they created.

1. Finiteness results in Diophantine geometry

The following conjecture was made by Serge Lang in [**L1**]:

Lang conjecture on subvarieties of Abelian varieties. *Let A be an Abelian variety over a field K of characteristic zero. Let $\Gamma \subset A(\overline{K})$ be a subgroup of finite rank and $X \subset A$ a closed subvariety. Then there exist in $X(\overline{K})$ finitely many translates of Abelian subvarieties whose union contains $X(\overline{K}) \cap \Gamma$.*

Here $\overline{K}$ denotes the algebraic closure of K. Also, recall that an Abelian variety is, by definition, a projective variety which has a structure of algebraic group. Finally, recall that the rank of a commutative group Γ is defined to be the dimension, as a $\mathbf{Q}$-vector space, of $\Gamma \otimes_{\mathbf{Z}} \mathbf{Q}$. Lang's conjecture was eventually proved, in the form above, through the work of Faltings [**F1, F2**] and Hindry [**H1**]; for a history of work on this conjecture (in particular for its relation with the Mordell conjecture), as well as for further developments we refer to [**H2**]. The first main purpose of these lectures is to explain the proof of a special case of this conjecture (the geometric case) following the author's paper [**B2**]; this proof is based on a method which has an interest in its own right and consists in replacing Γ in Lang's conjecture by a bigger subgroup that becomes a geometric object in a richer geometry, the so-called *differential algebraic geometry.* This idea triggered a remarkable new interaction between mathematical logic and Diophantine geometry; (cf. Hrushovski's proof of Lang's conjecture over function fields of characteristic p [**H3**]).

The first main result that we would like to explain is:

THEOREM (Geometric Lang conjecture [**B2**]). *Lang's conjecture is true in case K is a function field over an algebraically closed field k and A has K/k-trace zero.*

Recall that A is said to have K/k-trace zero if A contains no nonzero Abelian subvariety that descends to k (i.e. $\overline{K}$-isomorphic to a variety defined over k.) Another purpose of these lectures is to explain, following [**B4**], how the differential algebraic method in [**B2**] can be pushed to work "beyond" the Lang conjecture above, becoming relevant to (the geometric case of) another remarkable conjecture, the so-called *abc* conjecture (due to Masser and Oesterlé, cf. [**L2**], and generalized by the author [**B4**] and Vojta).

To explain this generalized "*abc* conjecture" we need some preparation.

Let K be a field equipped with a family of absolute values $| \ |_v \colon K \to [0, \infty)$, $v \in M$, all of which, except finitely many, are nonarchimedian. Set $v(x) = -\log |x|_v$ for $x \in K^{\times}$. Assume (m_v) is a collection of positive integers such that the "product formula"

$$\sum_v m_v v(x) = 0, \quad x \in K^{\times}$$

holds. Set

$$\gamma_v := \inf\{v(K^{\times}) \cap (0, \infty)\}.$$

Also, for any $\eta = (\eta_1, \dots, \eta_N) \in K^N$ and $v \in M$ set

$$v(\eta) = \min_j v(\eta_j).$$

Define the (affine, logarithmic) height

$$\text{height}_{\mathbf{A}^N} : \mathbf{A}^N(K) = K^N \to [0, \infty)$$

by the formula

$$\text{height}_{\mathbf{A}^N}(\eta) = -\sum_{v(\eta) \leq 0} m_v v(\eta) = \sum_v m_v \max_j \log^+ |\eta_j|_v$$

where $\log^+ x := \max\{\log x, 0\}$, $x \in [0, \infty)$. Note that

$$\text{height}_{\mathbf{A}^N}(\eta) = \text{height}_{\mathbf{P}^N}(1 : \eta_1 : \dots : \eta_N)$$

where

$$\mathrm{height}_{\mathbf{P}^N}(x_0 : \cdots : x_N) = \sum_v m_v \max_j \log |x_j|_v$$

is the usual height in projective space. On the other hand define the (logarithmic) conductor

$$\mathrm{cond}_{\mathbf{A}^N} : \mathbf{A}^N(K) = K^N \to [0, \infty)$$

by the formula

$$\mathrm{cond}_{\mathbf{A}^N}(\eta) = \sum_{v(\eta) \leq 0} m_v \gamma_v.$$

Clearly, by the very definition of γ_v we have

$$\mathrm{cond}_{\mathbf{A}^N}(\eta) \leq \mathrm{height}_{\mathbf{A}^N}(\eta), \quad \eta \in K^N.$$

We need the following piece of notation. Let $f, g\colon S \to [0, \infty)$ be two real functions on a set S; we write

$$f \leq g + O(1)$$

if there exists a positive real constant C such that

$$f(P) \leq g(P) + C, \quad P \in S$$

and we write $f = g + O(1)$ if $f \leq g + O(1)$ and $g \leq f + O(1)$. We write

$$f \ll g + O(1)$$

if there exist two real positive constants C_1, C_2 such that

$$f(P) \leq C_1 g(P) + C_2, \quad P \in S$$

and we write $f \equiv g + O(1)$ if $f \ll g + O(1)$ and $g \ll f + O(1)$. Coming back to *height* and *cond* one easily checks that if $P\colon \mathbf{A}^N(K) \to \mathbf{A}^n(K)$ is a map given by an n-tuple of polynomials in N variables with K-coefficients then we have

$$\mathrm{height}_{\mathbf{A}^n} \circ P \ll \mathrm{height}_{\mathbf{A}^N} + O(1),$$
$$\mathrm{cond}_{\mathbf{A}^n} \circ P \leq \mathrm{cond}_{\mathbf{A}^N} + O(1).$$

This permits one to define the **height** and the **conductor** for any affine variety as follows. Let U be an affine variety over K. Let $i\colon U \to \mathbf{A}^N$ be a closed immersion and define

$$\mathrm{height}_U\colon U(K) \to [0, \infty),$$
$$\mathrm{cond}_U\colon U(K) \to [0, \infty)$$

by the formulae

$$\mathrm{height}_U(P) := \mathrm{height}_{\mathbf{A}^N}\big(i(P)\big), \quad P \in U(K),$$
$$\mathrm{cond}_U(P) := \mathrm{cond}_{\mathbf{A}^N}\big(i(P)\big), \quad P \in U(K).$$

By the above discussion, if height_U and cond_U are defined by a closed immersion i and height'_U and cond'_U are defined by a closed immersion i' then

$$\mathrm{height}_U \equiv \mathrm{height}'_U + O(1),$$
$$\mathrm{cond}_U = \mathrm{cond}'_U + O(1).$$

In particular we have

$$\mathrm{cond}_U \ll \mathrm{height}_U + O(1).$$

Here is our basic definition.

DEFINITION. We say that the *abc* estimate holds on U if

$$\text{height}_U \ll \text{cond}_U + O(1).$$

We would like to understand what are, conjecturally, the affine varieties U on which the *abc* estimate holds. To tackle this question, and establish the link with the "usual" *abc* we need to be more specific about our field K. Assume, in what follows, that we are in one of the following situations:

1.1. Number field case. K is a number field equipped with its standard family of absolute values $(|\ |_v)$ (normalized in such a way that they extend the standard absolute values of $\mathbf{Q}$; in particular, if v divides a rational prime p, then $|p|_v = p^{-1}$.) We take $m_v = [K_v : \mathbf{Q}_v]$. So, for any nonarchimedian v, dividing an unramified rational prime p, we have $\gamma_v = v(p) = \log p$ so $m_v\gamma_v = \log \mathbf{N}v$ where $\mathbf{N}v$ is the norm of v (cardinality of the residue field of v). Since there are only finitely many ramified primes, we have

$$\text{cond}_{\mathbf{A}^N}(\eta) = \sum_{v(\eta)<0} \log \mathbf{N}v, \quad \eta \in K^N.$$

1.2. Function field case. K is a function field of one variable over an algebraically closed field k of characteristic zero. We equip K with the absolute values $|\ |_v$ arising from the k-rational points v of the smooth projective model V of K/k; we normalize them via the condition $\gamma_v = 1$ for all v, and we take $m_v = 1$ for all v. So in this case $\text{cond}_{\mathbf{A}^N}(\eta)$, for $\eta \in K^N$, is simply the number of points $v \in V$ which are poles for at least one of the rational functions η_j.

It is a trivial exercise to show that if, say, $K = \mathbf{Q}$ or $K = k(t)$, and if there exists a nonconstant morphism of K-varieties

$$\mathbf{P}^1 \backslash \{0, \infty\} \to U$$

into an affine K-variety U then the *abc* estimate fails on U. On the other hand the optimist would be tempted to believe that the presence of such morphisms is the only obstruction to the *abc* estimate; we make, optimistically, the following:

CONJECTURE. *Assume U is a smooth affine variety over K and assume that any morphism of K-varieties $\mathbf{P}^1 \backslash \{0, \infty\} \to U$ is constant. Then the abc estimate holds on U.*

Our conjecture is a consequence of the general "Vojta conjectures" (as shown by Vojta) and should be viewed as an affine analogue of the following:

LANG CONJECTURE ON VARIETIES OF GENERAL TYPE ([**L2**]). *If X is a smooth projective variety over K and if any morphism from an algebraic group G to X is constant then the points of $X(K)$ have bounded height; hence, in the number field case, $X(K)$ is finite.*

In the above conjecture of Lang one allows morphisms $G \to X$ defined over the algebraic closure $\overline{K}$ of K; it might be reasonable to allow this in our conjecture as well. In case $K = \mathbf{Q}$ and $U = \mathbf{P}^1 \backslash \{0, 1, \infty\}$, the conjecture above simply says that the *abc* estimate holds on this particular U; note that this is equivalent to the following:

THEOREM (Variant of the *abc* conjecture of Masser – Oesterlé). *For any relatively prime integers a, b, c with $a+b=c$ there exist real numbers $C>1$ and $\mu>1$ such that*

$$\max\{|a|,|b|,|c|\} \leq C \cdot \mathrm{rad}(abc)^{\mu}.$$

As usual $\mathrm{rad}(n)$, where n is an integer, is defined as the product of all primes dividing n; this variant is weaker than the *abc* of Masser and Oesterlé which predicts that μ can be made as close to 1 as we want. However this variant still implies, say, the asymptotic Fermat. To see the equivalence between the conjecture and the variant of *abc* for $K=\mathbf{Q}$ and $U=\mathbf{P}^1\backslash\{0,1,\infty\}$, embed U into $\mathbf{A}^2$ via the map

$$(1:x) \mapsto \left(x, \frac{1}{x(x-1)}\right).$$

Then, for coprime integers a, b, c with $a+b=c$, we have that $\mathrm{height}_U(1\colon c/a)$ equals

$$\mathrm{height}_{\mathbf{P}^2}\left(1:\frac{c}{a}:\frac{a^2}{cb}\right) = \log\max(|abc|,|b^2c|,|a^3|) \equiv \log\max(|a|,|b|,|c|) + O(1).$$

On the other hand

$$\mathrm{cond}_U\left(1:\frac{c}{a}\right) = \mathrm{cond}_{\mathbf{A}^2}\left(\frac{c}{a},\frac{a^2}{cb}\right) = \log\mathrm{rad}(abc).$$

Note that, by a result due (independently) to Mason and Silverman, the *abc* estimate holds, in the function field case, for the projective line minus 3 points. On the other hand results of Voloch [**V**], Brownawell – Masser [**BM**], and Wang [**W**] can suitably be interpreted as *abc* estimates, over function fields, "outside some exceptional loci" for projective spaces minus unions of hyperplanes.

THEOREM (*abc* for Abelian varieties with trace zero [**B4**]). *In the function field case the abc estimate holds for any affine open set of an Abelian variety with trace zero.*

THEOREM (*abc* for isotrivial Abelian varieties [**B8**]). *Let K be a function field over k, let A_k be an Abelian variety over k and D_k a divisor in A which does not contain any translate of a nonzero Abelian subvariety (in particular D_k is ample so $U_k := A_k\backslash D_k$ is affine). Let $U := U_k \otimes_k K$. Then the abc estimate holds on U.*

The above theorems are immediate consequences of the following stronger results:

THEOREM (Bounded multiplicity theorem, trace zero case [**B4**]). *Assume A/K is an Abelian variety with trace zero and let $f\in K(A)$ be a rational function. Then there exists a constant C depending only on K, A, f with the following property. For any point $P\in A(K)$ where f is defined and does not vanish, all zeroes and poles of $f(P)\in K^\times$ have multiplicity at most C.*

THEOREM (Bounded multiplicity theorem, isotrivial case [**B8**]). *Let X be a smooth projective curve over k, A an Abelian variety over k, and D an effective divisor on A. Assume that D contains no translate of a nonzero Abelian subvariety. Then there exists a real constant $C>0$, depending only on X, A, and D with the property that for any morphism $f\colon X\to A$, with $f(X)\not\subset D$, all points of the divisor f^*D have multiplicity at most C.*

A characteristic p version of these results was proved by Scanlon [**S1**].

2. Finiteness results in differential algebraic geometry

We begin by reviewing the basic concepts of differential algebra, following Ritt and Kolchin [**K**]. Roughly speaking, differential algebra is devoted to the study of algebraic (nonlinear) differential equations in the same way in which algebra is devoted to the study of algebraic equations. So, as a principle, analysis is banned from the foundations of differential algebra. Nevertheless, many proofs in differential algebra (in particular some of the proofs of our theorems) as well as many applications of differential algebra are analytic. (For the role of differential algebra in the study of nonoscillatory properties of solutions of algebraic differential equations, for instance, we refer to Lou van den Dries' work on this subject. For a survey on o-minimal structures see [**vdD**].)

To introduce the main concepts of differential algebra we start with a field $\mathcal{F}$ of characteristic zero equipped with a derivation δ (i.e. an map $\delta\colon \mathcal{F} \to \mathcal{F}$ that satisfies the Leibnitz rule.) The pair $(\mathcal{F}, \delta)$ will be referred to as a δ-field. We let $\mathcal{C}$ be the constant field, i.e. the field of all $x \in \mathcal{F}$ such that $\delta x = 0$. (If K is a function field we will always assume we have fixed a nonzero k-derivation on it and (K, d) is embedded into $(\mathcal{F}, \delta)$.) One defines the δ-polynomial ring $\mathcal{F}\{T\}$, where T is an n-tuple of variables, as the (usual) polynomial ring over $\mathcal{F}$ in the variables $T^{(i)}$, $i \geq 0$. The *order* of $A \in \mathcal{F}\{T\}$ is the highest i such that a variable $T_j^{(i)}$ is present in A. $\mathcal{F}$ is called δ-closed if for any $A, B \in \mathcal{F}\{T\}$, T a variable, such that $\operatorname{ord} B < \operatorname{ord} A$, there exists $a \in \mathcal{F}$ such that $A(a) = 0$ and $B(a) \neq 0$. Any δ-field can be embedded into a δ-closed δ-field.

We need more definitions (cf. our papers [**B2, B3, B4, B5, B8**].). A D-scheme is an $\mathcal{F}$-scheme V with a given derivation on $\mathcal{O}_V$ that lifts δ. D-schemes form a category (morphisms are required to commute with the derivations). A D-group scheme is a D-scheme which is also an $\mathcal{F}$-group scheme such that the multiplication, inverse, and unit are morphisms of D-schemes. A D-variety is a D-scheme which is also a variety over $\mathcal{F}$. An algebraic D-group is a D-group scheme which is also an algebraic group over $\mathcal{F}$. One has a forgetful functor

$$\{D\text{-schemes}\} \to \{\mathcal{F}\text{-schemes}\}, \quad V \mapsto V^!.$$

It has a right adjoint

$$\{\mathcal{F}\text{-schemes}\} \to \{D\text{-schemes}\}, \quad X \mapsto X^\infty,$$
$$\operatorname{Hom}_{\mathcal{F}}(V^!, X) = \operatorname{Hom}_D(V, X^\infty)$$

defined as follows. If $X = \operatorname{Spec} \mathcal{F}[T]/I$, set $X^\infty := \operatorname{Spec} \mathcal{F}\{T\}/[I]$, $[I] := (I, \delta I, \dots)$. In the nonaffine case one glues the affine pieces. (Note that X^∞ is the inverse limit of a system X^n of varieties obtained by truncating everything to order n. One has a natural map $\nabla\colon X(\mathcal{F}) \to X^\infty(\mathcal{F})$ which in coordinates sends $a \in \mathcal{F}$ into $(a, \delta a, \delta^2 a, \dots)$. Pullbacks via this map of Zariski closed sets are called δ-closed sets. If $\mathcal{F}$ is δ-closed there is a bijection between delta closed sets and reduced closed D-subschemes of X^∞. If $\Sigma \subset X(\mathcal{F})$ is δ-closed, corresponding to a D-subscheme $H \subset X^\infty$ then one defines the absolute dimension $a(\Sigma)$ as the maximum of the transcendence degrees over $\mathcal{F}$ of the function fields of the irreducible components of H.

EXAMPLE. $a\big(X(\mathcal{F})\big) = \infty$ if $\dim X > 0$. If X descends to $\mathcal{C}$ (i.e. comes from a variety $X_{\mathcal{C}}$ over $\mathcal{C}$) then $a\big(X_{\mathcal{C}}(\mathcal{C})\big) = \dim X$. For G an algebraic group over $\mathcal{F}$, the above bijection induces a bijection between the δ-closed subgroups of finite absolute dimension of $G(\mathcal{F})$ and algebraic D-subgroups of G^∞. By a δ-function on $X(\mathcal{F})$, where X is a variety, we mean a map $X(\mathcal{F}) \to \mathcal{F}$ obtained by composing ∇ with a regular map on X^∞. A δ-character on an algebraic group will mean a δ-function which is also an additive homomorphism. A δ-map $X \to \mathbf{A}^n$ will mean a map $X(\mathcal{F}) \to \mathcal{F}$ whose components are δ-functions.

The next theorem about δ-functions on unirational varieties will not be needed for our Diophantine results on Abelian varieties; but it is quite instructive to compare the situation of Abelian varieties (to be considered below) with that of unirational varieties. (Recall that a variety X is called unirational if there is a dominant rational map $\mathbf{A}^n \to X$.)

THEOREM (δ-density theorem [**B3**]). *If X is a smooth projective unirational $\mathcal{F}$-variety then any δ-function on $X(\mathcal{F})$ is constant. Moreover, if X is defined over an intermediate field L between $\mathcal{C}$ and $\mathcal{F}$, $L \neq \mathcal{C}$, then $X(L)$ is δ-dense in $X(\mathcal{F})$.*

Unlike the previous theorem, the next theorem will play a key role in our Diophantine applications.

THEOREM (Finiteness of absolute dimension [**B2**]). *Let A be an Abelian $\mathcal{F}$-variety of dimension g. Then the intersection $A^\sharp$ of the kernels of all δ-characters of A has finite absolute dimension $g \leq a(A^\sharp) \leq 2g$. Consequently, the δ-closure of any finite rank subgroup of $A(\mathcal{F})$ has finite absolute dimension $\leq 2g + r$, where r is the rank. In particular, in the trace zero case, the δ-closure of the group of division points of $A(K)$ has finite absolute dimension.*

Recall that the group of division points of $A(K)$ is, by definition, the group of all points in $A(\overline{K})$ that have a multiple in $A(K)$.

The δ-characters are the incarnation, in differential algebraic geometry, of the "Manin homomorphisms"; but our finiteness result $\left(a(A^\sharp) < \infty\right)$ is quite different in nature from the Manin – Chai "theorem of the kernel". What the latter says, in the trace zero case, in our terminology is the following:

THEOREM OF THE KERNEL (Manin – Chai). *Let A be an Abelian $\mathcal{F}$-variety with $\mathcal{F}/\mathcal{C}$-trace zero. Then for any intermediate δ-field $\mathcal{C} \subset L \subset \mathcal{F}$ of definition for A which is finitely generated over $\mathcal{C}$ we have $A^\sharp \cap A(L) = A(L)_{\text{tors}}$.*

We include one more result which is not directly related to our Diophantine applications but completes the picture, emerging from the above statements, as to the behaviour of (classical) algebraic varieties from the viewpoint of differential algebraic geometry.

THEOREM (δ-maps on curves [**B3**]). *Let X be a smooth projective curve over $\mathcal{F}$ of genus $g \geq 2$, that does not descend to $\mathcal{C}$. Then there exists an injective δ-map $\phi \colon X \to \mathbf{A}^n$.*

In a certain precise sense one can actually prove more, namely that projective curves of genus at least 2 are affine in this geometry. Halfway towards our *abc* for Abelian varieties we have the following differential algebraic generalization of the geometric Lang conjecture on subvarieties of Abelian varieties:

THEOREM (Differential algebraic Lang [**B2**]). *Let A be an Abelian variety over $\mathcal{F}$ with trace zero over $\mathcal{C}$. Let $\Sigma \subset A(\mathcal{F})$ be a δ-closed subgroup of finite absolute* *dimension and $X \subset A$ a closed subvariety. Then there exist in X finitely many translates of Abelian subvarieties whose union contains $X(\mathcal{F}) \cap \Sigma$.*

The above theorem is indeed a generalization of the geometric Lang conjecture because, together with the theorem on finite absolute dimension, it trivially implies the geometric Lang conjecture stated in Section 1.

3. Discussion of the proofs

3.1. Descent results. An $\mathcal{F}$-variety (resp. an algebraic group over $\mathcal{F}$) is said to descend to constants if it comes, via base change, from a variety (resp. an algebraic group) over $\mathcal{C}$.

The next four theorems are proved via complex analytic arguments; for the first theorem Henri Gillet showed me an argument based on formal schemes; this kind of argument, from formal geometry, does not seem to apply to the other results.

THEOREM (Descent of projective varieties [**B1**]). *Any projective D-variety descends to constants.*

THEOREM (Descent of linear algebraic groups [**B2**]). *Any linear algebraic D-group descends to constants.*

Let us remark that the above fails for nonlinear groups. A very interesting question is: let G be an algebraic D-group and H be its maximum connected linear algebraic subgroup; does H descend to constants?

We won't need the previous two theorems for our Diophantine purposes. For the differential algebraic Lang theorem and for the *abc* theorem for Abelian varieties with trace zero we will need:

THEOREM (Descent via D-groups [**B2**]). *Let G be an algebraic D-group, $V \subset G$ a closed D-subvariety, and $V \to W$ a dominant morphism to a projective variety W of general type. Then the Albanese variety* $\mathrm{Alb}(W)$ *descends to constants.*

For the *abc* theorem for isotrivial Abelian varieties we will need:

THEOREM (Descent via D-groups, split case [**B8**]). *Let W be a projective variety of general type over K. Assume W is a closed subvariety of A_K, where A is an Abelian k-variety. Let G be any algebraic D-group, $V \subset G$ a D-subvariety and $u\colon V \to W$ be a dominant morphism. Then, after replacing K by a finite extension of it, one may find a closed k-subvariety $Z \subset A$ and a point $Q \in A(K)$ such that $W = Z_K + Q$ in A_K. Moreover, if we view W as a D-scheme by trivially lifting δ from K to $W \simeq Z_K = Z \otimes_k K$, then $u\colon V \to W$ is necessarily a morphism of D-schemes.*

Let us sketch the proof of the theorem on "descent via D-groups". We may assume $\mathcal{C}$ is the field of complex numbers. We may further reduce ourselves to the case when $(V, W, G, \operatorname{Spec} \mathcal{F}, \delta)$ comes, via base change, from a "situation in complex algebraic geometry" $(\mathcal{V}, \mathbf{W}, \Gamma, \mathbf{S}, \delta_{\mathbf{S}})$ where

- $\mathbf{S}$ is an affine complex curve with a nowhere vanishing vector field $\delta_{\mathbf{S}}$.
- $\mathbf{W}$ is projective over $\mathbf{S}$ with integral geometric fibers of general type.
- Γ is a group scheme of finite type over $\mathbf{S}$ equipped with a vector field δ_Γ, lifting $\delta_{\mathbf{S}}$, such that the inverse and the multiplication on Γ are δ_Γ-equivariant (call such a δ_Γ "group compatible").

$\mathcal{V}$ is a closed subvariety of Γ, horizontal with respect to δ_Γ, with a dominant map to $\mathbf{W}$.

A lemma of Hamm says that an "analytic 1-parameter family of complex Lie groups" whose total space is equipped with a "group-compatible" analytic vector field lifting a nonvanishing vector field on the base is locally analytically trivial. So locally analytically $\Gamma = \Gamma_0 \times \mathbf{S}$, $\delta_\Gamma = (0, \delta_{\mathbf{S}})$, where Γ_0 is some Lie group. (So we have an analytic trivialization; note that there is no algebraic trivialization in general!) It follows that $\mathcal{V} \to \mathbf{S}$ is locally analytically trivial. Fix $s_0 \in \mathbf{S}$. For any $s \in \mathbf{S}$ consider the analytic map

$$\phi_s \colon \mathcal{V}_{s_0} \simeq \mathcal{V}_s \to \mathbf{W}_s.$$

Its image contains a Zariski open set. The big Picard theorem [**KO**] says that any analytic map from an algebraic variety to a projective variety of general type, whose image contains a complex open set, is in fact algebraic. So our maps ϕ_s are algebraic. So they induce maps from the Albanese variety of (a smooth projective model of) $\mathcal{V}_{s_0}$ into $\mathrm{Alb}(\mathbf{W}_s)$. So all $\mathrm{Alb}(\mathbf{W}_s)$ are isomorphic to each other. This implies that $\mathrm{Alb}(W)$ descends to constants and we are done.

3.2. Sketch of proof of the finite absolute dimension theorem. Set $C = \mathrm{Spec}\,\mathcal{O}(A^\infty)$. One easily checks that $U := \mathrm{Ker}(A^\infty \to A)$ is unipotent. Then one proves that whenever one has an extension

$$0 \to U \to G \to A$$

with A an Abelian variety and U unipotent (infinite dimensional) then one also has an exact sequence

$$0 \to \tilde{A} \to G \to \tilde{U} \to 0$$

with $\tilde{A}$ finite dimensional and $\tilde{U}$ unipotent (infinite dimensional). (This is an exercise in the theory of proalgebraic groups.) Then one concludes by noting that $\tilde{A}$ can be chosen to be a D-subgroup and that $A^\sharp$ is the δ-closed subset corresponding to $\tilde{A}$.

3.3. Proof of differential algebraic Lang. Assume we are in the hypothesis of differential algebraic Lang. Then Σ corresponds to an algebraic D-subgroup G of A^∞. Let V be an irreducible component of $X^\infty \cap G$ and let $Y \subset X$ be the Zariski closure of the image of V in X. We claim Y is a translate of an Abelian subvariety, and we shall be done. Assume it's not. By a result of Ueno [**U**, pp. 117, 120], there is an Abelian subvariety $A_1 \subset A$ such that the image W of Y in A/A_1 is positive dimensional, of general type. By the theorem on "descent via D-groups" it follows that $\mathrm{Alb}(W)$ descends to constants. But A was assumed to have trace zero, a contradiction.

3.4. Proof of *abc* for Abelian varieties.

Preparation. Let A be an Abelian variety over K, with K/k-trace zero and let U be an affine open set. The closed set $A \backslash U$ is the support of a very ample effective divisor D on A. Note that D is neither irreducible nor reduced a priori. Embed A into $\mathbf{P}^N_K$ such that D is given by $x_0 = 0$, where $x_0, \ldots, x_N$ are a basis of $H^0(\mathbf{P}^N_K, \mathcal{O}(1))$. Let $U_j \subset A$ be the open sets defined by $x_j \neq 0$. So $U_0 = U$. We need to show that there is a constant C such that for any point $\eta = (\eta_1, \ldots, \eta_N) \in U(K)$ and for any place v we have $v(\eta) \geq C$. This actually easily implies the stronger bounded multiplicity theorem as well.

For simplicity we shall assume A is simple.

Constructing the finite set Y. Let A^∞, D^∞ be attached to A and D respectively (the whole construction being made over K, rather than over $\mathcal{F}$). Then D^∞ appears as a closed subscheme of A^∞. Exactly as in the proof of Differential Algebraic Lang, there exists an algebraic D-subgroup $G \subset A^\infty$ such that for all $P \in A(K)$ we have $\nabla(P) \in G(K)$. Set $V := D^\infty \cap G$ (scheme theoretic intersection) and let $Y \subset A$ be the Zariski closure of the image of V via the map $A^\infty \to A$; we view Y with its structure of reduced subscheme of A. As in the proof of differential algebraic Lang Y is isomorphic, over the algebraic closure of K, to a finite union of translates of Abelian subvarieties. Since A was assumed simple, Y is finite.

For simplicity we shall assume the finitely many points of Y are rational over K.

"*Uniform discreteness*": $\boldsymbol{\phi_{ij}}$ *and* $\boldsymbol{\gamma}$. It is an easy consequence of the Manin–Chai theorem of the Kernel that one can find, for each j, a finite family $(\phi_{ij})_{i\in I_j}$, $\phi_{ij} \in \mathcal{J}(Y \cap U_j)$ and there exists a real number $\gamma \geq 0$ such that for any v and any $P \in U_j(K)\backslash Y$, there exists an index $i \in I_j$ such that $v\big(\phi_{ij}(P)\big) \leq \gamma$. (Here $\mathcal{J}$ of a subscheme denotes the ideal defining that subscheme.) Identify $\phi_{ij} \in \mathcal{O}(U_j)$ with their pull-backs in $\mathcal{O}(U_j^\infty)$. Then, for a suitable q, we have

$$\phi_{ij}^q \in \mathcal{J}(V \cap U_j^\infty) = \mathcal{J}(D^\infty \cap U_j^\infty) + \mathcal{J}(G \cap U_j^\infty).$$

Clearly

$$\mathcal{J}(D \cap U_j) = \frac{x_0}{x_j} \cdot \mathcal{O}(U_j).$$

So we get

$$\mathcal{O}(D^\infty \cap U_j^\infty) = \left(\frac{x_0}{x_j}, \delta\left(\frac{x_0}{x_j}\right), \delta^2\left(\frac{x_0}{x_j}\right), \dots\right)\mathcal{O}(U_j^n).$$

We may write, for all j and $i \in I_j$:

$$\phi_{ij}^q = \sum_{t=0}^{n} F_{ijt}\left(\frac{x_0}{x_j}, \dots, \frac{x_N}{x_j}\right)\delta^t\left(\frac{x_0}{x_j}\right) + g_{ij} \tag{1}$$

with $g_{ij} \in \mathcal{J}(G \cap U_j^\infty) \subset \mathcal{O}(U_j^\infty)$ and F_{ijt} differential polynomials with K-coefficients.

Defining β. Clearly we may find a real number $\beta \geq 0$ such that for any $\eta \in K^N$ and any v with $v(\eta) \geq 0$ we have $v\big(F_{ijt}(\eta)\big) \geq -\beta$ for all i, j, t.

Fixing P *and* v. Now we fix a point $P \in U(K) = A(K)\backslash D$ with coordinates $\xi \in K^{N+1}$ and fix a place v. Estimate equation (1) at $\nabla(P) \in A^\infty(K)$. Since $\nabla(P) \in G(K)$ we have $g_{ij}\big(\nabla(P)\big) = 0$. So by further taking v we get, for each j, and each $i \in I_j$,:

$$q \cdot v(\phi_{ij}(P)) \geq \min_t\left\{v\left(F_{ijt}\left(\frac{\xi_0}{\xi_j}, \dots, \frac{\xi_N}{\xi_j}\right)\right)\right\} + \min_t\left\{v\left(\delta^t\left(\frac{\xi_0}{\xi_j}\right)\right)\right\}. \tag{2}$$

Choosing j. Now for our fixed P and v there exists j such that $\xi_j \neq 0$ (hence $P \in U_j(K)$) and $v(\xi_i/\xi_j) \geq 0$ for all i, i.e. $v(\xi_j) = v(\xi)$. Fix such a j.

Choosing i. For our fixed P, v, j there exists, by "uniform discreteness", an index $i \in I_j$ such that

$$v\big(\phi_{ij}(P)\big) \leq \gamma. \tag{3}$$

Fix such an index i.

Conclusion

Now, by the choice of β, we have, for all t,

$$v\left(F_{ijt}\left(\frac{\xi_0}{\xi_j},\dots,\frac{\xi_N}{\xi_j}\right)\right) \geq -\beta. \tag{4}$$

It is trivial to see that there exists a real number $\theta \geq 0$ such that for all $x \in K$ and all v we have $v(\delta x) \geq v(x) - \theta$. Then putting together (2), (3), and (4), we get

$$q \cdot \gamma \geq q \cdot v(\phi_{ij}(P)) \geq -\beta + (v(\xi_0) - v(\xi_j) - n \cdot \theta).$$

If $\eta = (\xi_1/\xi_0, \dots, \xi_N/\xi_0)$ then we get

$$v(\eta) = v(\xi) - v(\xi_0) \geq C := -(\beta + q \cdot \gamma + n \cdot \theta)$$

and we are done.

3.5. Link with the Painlevé property. We conclude our notes by drawing the attention to an interesting aspect of our proof of the theorem on "descent via D-groups", which is actually the key technical point of our whole approach. In the proof of that theorem we were able to exploit the interaction between big Picard and a certain remarkable property of the foliation defined by δ_Γ on Γ. Let us isolate this property and comment upon it. Assume we are given the following data:

$$(f\colon \mathbf{X} \to \mathbf{S}, \delta_{\mathbf{S}}, \delta_{\mathbf{X}})$$

where $\mathbf{S}$ is an affine complex algebraic curve with a nowhere vanishing vector field $\delta_{\mathbf{S}}$, $\mathbf{X}$ is a complex algebraic manifold equipped with a vector field $\delta_{\mathbf{X}}$, and f is a regular map of algebraic varieties which is horizontal with respect to $\delta_{\mathbf{X}}$ and $\delta_{\mathbf{S}}$.

Let us say that $(f\colon \mathbf{X} \to \mathbf{S}, \delta_{\mathbf{S}}, \delta_{\mathbf{X}})$ has *property* P if for any point $s_0 \in \mathbf{S}$ there exists an **analytic** neighbourhood $\mathbf{D} \subset \mathbf{S}$ of s_0 and an **analytic** isomorphism

$$\sigma\colon f^{-1}(\mathbf{D}) \to f^{-1}(s_0) \times \mathbf{D}$$

over $\mathbf{D}$ such that σ is horizontal with respect to the vector fields

$$\delta_{\mathbf{X}}|_{f^{-1}(\mathbf{D})} \quad \text{and} \quad 0 \times \delta_{\mathbf{S}}.$$

The main point of this definition is that our data $(f\colon \mathbf{X} \to \mathbf{S}, \delta_{\mathbf{S}}, \delta_{\mathbf{X}})$ are **algebraic** but we only require that there exist **analytic** "local trivialization" around each point. The "P" in "property P" stands for "Painlevé"; indeed if property P holds then the solutions of the algebraic differential equation that corresponds to our data can be analytically continued along any path on $\mathbf{S}$. This is a property similar to (but a priori stronger than) that of the usual Painlevé equations; indeed the latter may have, a priori, solutions with "movable poles" in $\mathbf{S}$. Whether or not the movable poles can be eliminated by replacing $\mathbf{X}$ with a variety birationally equivalent to it is not clear to us; an indication that this might often be possible is found in Okamoto's work [**O**]. In any case, as we saw in the proof of our theorem on "descent via D-groups", property P holds for the situation $(\Gamma \to \mathbf{S}, \delta_{\mathbf{S}}, \delta_\Gamma)$ in that proof (due to Hamm's result) and this, rather than the group scheme structure of $\Gamma/\mathbf{S}$ was what we have used. This suggests that property P could be a good substitute for the underlying group structure in Lang's Diophantine conjectures, at least in the geometric setting. At this point this is mere speculation but we think the above mentioned possibility deserves to be explored.

References

[BM] D. Brownawell and D. Masser, *Vanishing sums in function fields*, Math. Proc. Cambridge Philos. Soc. **100** (1986), 427–434.

[B1] A. Buium, *Fields of definition of algebraic varieties in characteristic zero*, Compositio Math. **61** (1987), 339–352.

[B2] ———, *Intersections in jet spaces and a conjecture of S. Lang*, Ann. of Math. (2) **136** (1992), 583–593.

[B3] ———, *Geometry of differential polynomial functions.* I: *Algebraic groups*, Amer. J. Math. **115** (1993), no. 6, 1385–1444; II: *Algebraic curves*, Amer. J. Math. **116** (1994), no. 4, 785–819; III: *Moduli spaces*, Amer. J. Math. **117** (1995), 1–73.

[B4] ———, *The abc theorem for Abelian varieties*, Internat. Math. Res. Notices (1994), no. 5, 219–233.

[B5] ———, *Differential algebra and Diophantine geometry*, Actualités Math., Hermann, Paris, 1994.

[B6] ———, *Differential characters of Abelian varieties over p-adic fields*, Invent. Math. **122** (1995), no. 2, 309–340.

[B7] ———, *Geometry of p-jets*, Duke Math. J. **82** (1996), no. 2, 349–367.

[B8] ———, *Intersection multiplicities on Abelian varieties*, Math. Ann. **310** (1998), no. 4, 653–659.

[vdD] L. van den Dries, *o-minimal structures and real analytic geometry*, Current Developments in Mathematics (Cambridge, MA, 1998), Int. Press, Somerville, MA, 1999, pp. 105–152.

[F1] G. Faltings, *Endlichkeitssätze für abelsche Varietäten über Zahlkörpern*, Invent. Math. **73** (1983), 349–366.

[F2] ———, *The general case of S. Lang's conjecture*, Barsotti Symposium in Algebraic Geometry (Abano Terme, 1991) (V. Cristante and W. Messing, eds.), Perspect. Math., vol. 15, Academic Press, San Diego, CA, 1994, pp. 175–182.

[H1] M. Hindry, *Autour d'une conjecture de Serge Lang*, Invent. Math. **94** (1988), no. 3, 575–605.

[H2] ———, *Introduction to Abelian varieties and the Mordell–Lang conjecture*, Model Theory and Algebraic Geometry (É. Bouscaren, ed.), Lecture Notes in Math., vol. 1696, Springer, Berlin, 1998, pp. 85–100.

[H3] E. Hrushovski, *The Mordell–Lang conjecture over function fields*, J. Amer. Math. Soc. **9** (1996), 667–690.

[KO] S. Kobayashi and T. Ochiai, *Meromorphic mappings into compact complex spaces of general type*, Invent. Math. **31** (1975), 7–16.

[K] E. Kolchin, *Differential algebra and algebraic groups*, Academic Press, New York, 1973.

[L1] S. Lang, *Division points on curves*, Ann. Mat. Pura Appl. (4) **70** (1965), 229–234.

[L2] ———, *Number theory* III, Springer, Heidelberg–New York, 1991.

[M1] Yu. I. Manin, *Rational points on an algebraic curve over function fields.*, Transl. Amer. Math. Soc. (2) **50** (1966), 189–234; Letter to the Editor, Izv. Akad. Nauk USSR **34** (1990), 465–466.

[M2] R. C. Mason, *Diophantine equations over function fields*, London Math. Soc. Lecture Notes Ser., vol. 96, Cambridge Univ. Press, Cambridge, 1984.

[O] K. Okamoto, *Sur les feuilletages associés aux equations du second ordre a points critiques fixes de P. Painlevé*, Japan J. Math. (N.S.) **5** (1979), no. 1, 1–79.

[S1] T. Scanlon, *The abc theorem for commutative algebraic groups in characteristic p*, Internat. Math. Res. Notices (1997), no. 18, 881–898.

[S2] J. H. Silverman, *The S-unit equation over function fields*, Math. Proc. Cambridge Philos. Soc. **95** (1984), no. 3, 3–4.

[U] K. Ueno, *Classification of algebraic varieties and compact complex spaces*, Lecture Notes in Math., vol. 439, Springer, Berlin, 1975.

[V] F. Voloch, *Diagonal equations over function fields*, Bol. Soc. Brasil. Mat. (N.S.) **16** (1985), 29–39.

[W] J. T. Y. Wang, *S-integral points of projective space omitting hyperplanes over function fields of positive characteristic*, (preprint).

APPENDIX A

Abstract: o-Minimal Structures, Real Analytic Geometry, and Transseries

L. VAN DEN DRIES

ABSTRACT. o-minimal structures provide a general framework for finiteness phenomena in real algebraic and real analytic geometry. "o-minimal" finiteness results usually come with bounds that are uniform in the parameters of the problem, which, of course, is suggestive in connection with Dulac's conjecture and Hilbert's 16th problem on limit cycles. There are also strong restrictions on the growth of functions definable in a given o-minimal structure, and on their asymptotic properties. A related but partly independent topic concerns transseries (Écalle, Van der Hoeven), also called "logarithmic-exponential series." No previous knowledge on these topics is assumed.

Lecture 1. o-minimal structures. I
Lecture 2. o-minimal structures. II
Lecture 3. Transseries. I
Lecture 4. Transseries. II

APPENDIX B

List of Lectures Delivered at the Workshop

A. Bolibrukh: *Asymptotic series and differential equations.* I – V

A. Buium: *Differential algebraic geometry and Diophantine geometry.* I – V

L. van den Dries:

- *o-minimal structures.* I
- *o-minimal structures.* II
- *Transseries.* I
- *Transseries.* II

F. Dumortier: *Hopf – Takens bifurcations and centers*

J.-P. Françoise: *New findings on Liénard equations*

Yu. Ilyashenko: *Hilbert type numbers for Abel and Liénard equations*

V. Kaloshin:

- *Hilbert – Arnol'd problem and an estimate for ciclicity of elementary polycycles (introduction)*
- *Normal forms and the Khovanskiĭ method*
- *Whitney stratification of algebraic varieties and an elementary proof of its existence*
- *A_P-stratification and its application to the composition*
- *Newton interpolation polynomials and applications to bifurcation of spacial limit cycles*

J. Llibre: *Darboux integrability and the inverse integrating factor*

C. Miller: *Differential equations over polynomially bounded o-minimal structures*

R. Moosa: *An effective finiteness theorem in differential algebraic geometry*

A. Mourtada: *Projection de sous-ensembles quasi-réguliers d'Hilbert.* I, II

J.-P. Rolin: *Élimination des quantificateurs pour les séries généralisées*

C. Rousseau: *Finite cyclycity of graphics of vector fields.* I, II

P. Speissegger: *Some non-analytic o-minimal structures*

S. Yakovenko:

- *Hilbert's* 16*th problem: Limit cycles, cyclicity, Abelian integrals*
- *Abelian integrals and differential equations*
- *Quasialgebraicity of function fields*
- *Quantitative theory of differential equations*
- *Isomonodromic reduction principle and the Riemann – Hilbert problem*

H. Zhu: *Finite cyclicity of graphics through a nilpotent singularity and applications to Hilbert's* 16*th problem for quadratic systems*

APPENDIX C

Photographs of Some Workshop Participants

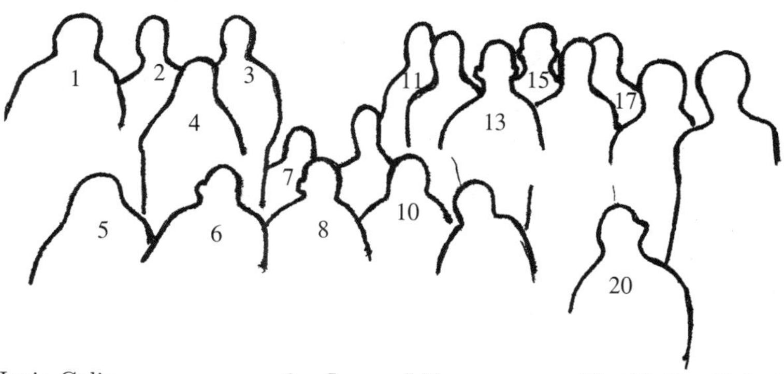

1	Iurie Calin	8	Jaume Llibre	15	Vadim Kaloshin
2	Dumitru Cozma	9	Christopher Miller	16	Luc Bélair
3	Patrick Speissegger	10	Sergei Yakovenko	17	Dmitry Novikov
4	Laura Farell	11	Jorge Pereira	18	Christian Michaux
5	Thierry Zell	12	Christiane Rousseau	19	Dana Schlomiuk
6	Nicolae Vulpe	13	Freddy Dumortier	20	Didier D'Acunto
7	Matthias Aschbrenner	14	Jean-Philippe Rolin		

From left to right: A. Buium, D. Schlomiuk, A. A. Bolibrukh

APPENDIX D

List of Participants

Matthias Aschenbrenner
University of Illinois at Urbana-Champaign, USA

Luc Bélair
Université du Québec à Montréal, Canada

Andrey Bolibrukh
Université de Strasbourg, France & Steklov Mathematical Institute, Russia

Alexandru Buium
University of New Mexico, USA

Iurie Calin
Academy of Sciences of Moldova, Moldova

Alina Carmen Cojocaru
Queen's University, Canada

Dumitru Cozma
Tiraspol State University, Moldova

Didier D'Acunto
Université de Savoie, France

Lou van den Dries
University of Illinois at Urbana-Champaign, USA

Freddy Dumortier
Limburgs Universitair Centrum, Belgium

Laura Farell
Université de Montréal, Canada

Jean-Pierre Françoise
Université Paris VI, France

Yulij Ilyashenko
Moscow Independent University & Steklov Mathematical Institute, Russia

Shafiqul Islam
Concordia University, Canada

Vadim Kaloshin
Princeton University, USA

The affiliations are the ones at the time the workshop took place.

Jaume Llibre
Universitat Autònoma de Barcelona, Spain

Christian Michaux
Université de Mons-Hainaut, Belgium

Christopher Miller
Ohio State University, USA

Rahim Moosa
University of Illinois at Urbana-Champaign, USA

Abderaouf Mourtada
Université de Bourgogne, France

Ram Murty
Queen's University, Canada

Dmitry Novikov
University of Toronto, Canada

Jorge Pereira
Instituto de Matematica Pura e Aplicada, Brazil

Wai Yan Pong
University of Illinois at Urbana-Champaign, USA

Jean-Pierre, Rolin
Université de Bourgogne, France

Christiane Rousseau
Université de Montréal, Canada

Dana Schlomiuk
Université de Montréal, Canada

Patrick Speissegger
University of Wisconsin-Madison, USA

Alexandru Suba
State University of Moldova, Moldova

Nicolae Vulpe
Academy of Sciences of Moldova, Moldova

Sergey Yakovenko
The Weizmann Institute of Science, Israel

Thierry Zell
Purdue University, USA

Huaiping Zhu
University of Waterloo, Canada